Field Notes

Field Notes

Richard Quinney

Borderland Books

Published by Borderland Books, Madison, WI
www.borderlandbooks.net

Publisher's Cataloging-in-Publication Data

Quinney, Richard.
Field Notes / Richard Quinney.—1st ed.
p. : ill. ; cm.
Photographs by the author.
Includes bibliographical references.

ISBN: 978-0-9768781-6-2

1. Nature. 2. Natural history. 3. Quinney, Richard—Homes and Haunts—Wisconsin—Walworth County. I. Title.

PS3567.U53 F54 2008
814.54 2007909720

Printed in the United States of America

First edition

The author of the following letters takes the liberty, with all proper deference, of laying before the public his idea of parochial history, which, he thinks, ought to consist of natural productions and occurrences as well as antiquities. He is also of the opinion that if stationary men would pay some attention to the districts on which they reside, and would publish their thoughts respecting the objects that surround them, from such materials might be drawn the most complete county-histories, which are still wanting in several parts of this kingdom, and in particular in the county of Southampton.

Gilbert White

The Natural History of Selborne

Contents

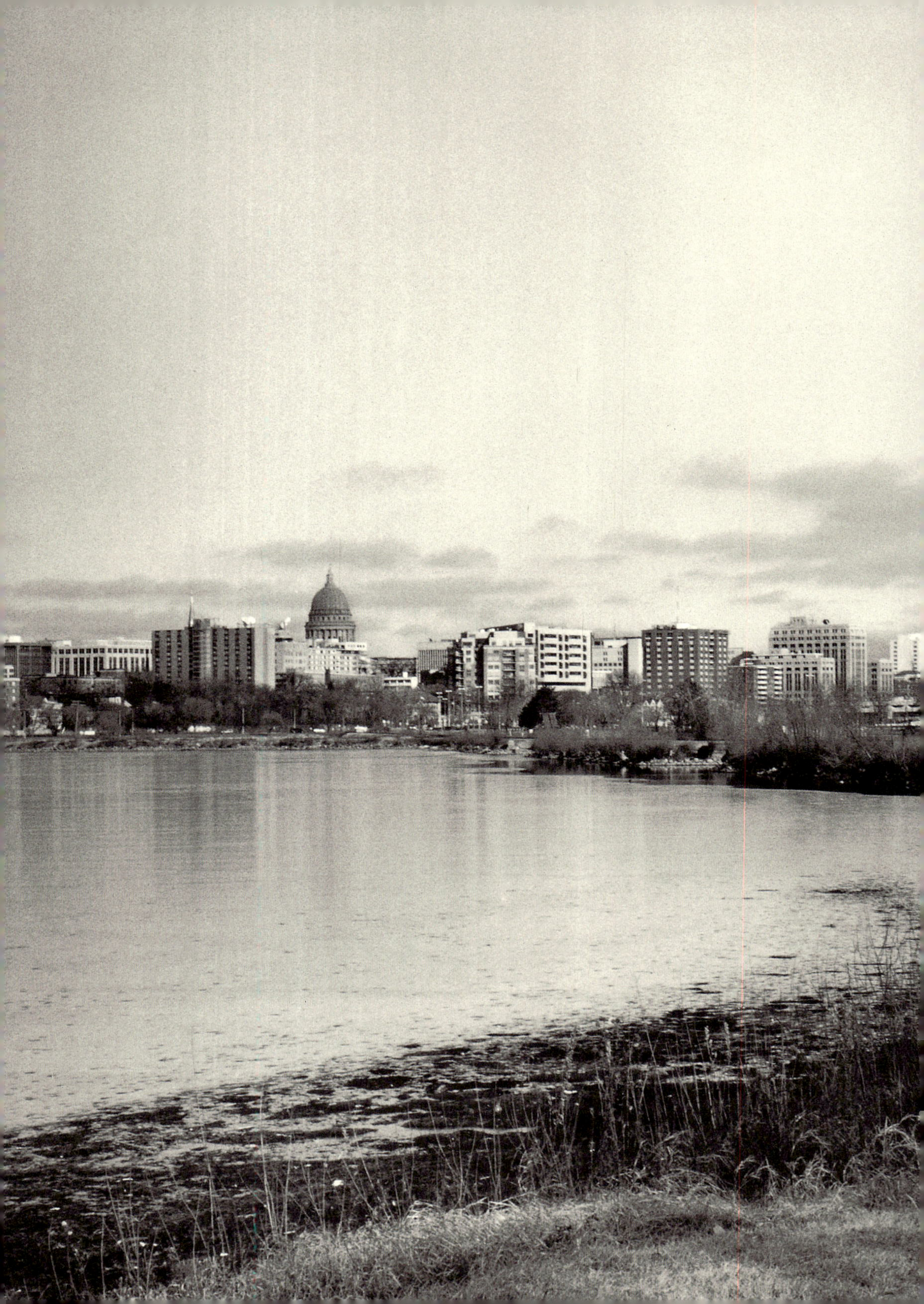

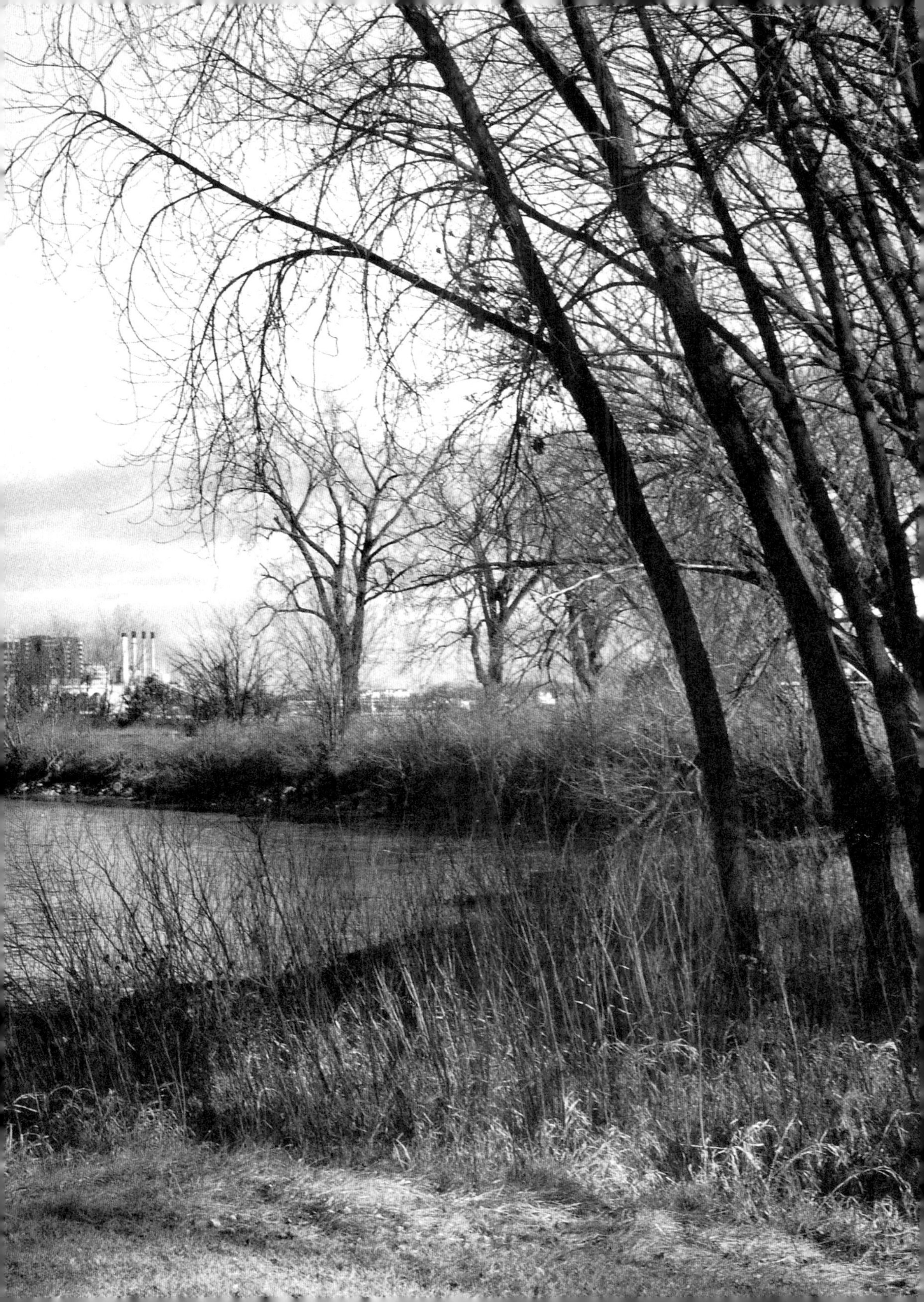

Prologue

Mystery of the Marsh

Once upon a time, on the day of his thirteenth birthday, the boy feared that soon he would die. All around him as he had grown up on the farm he had seen life come and go. Plants sprang from the earth each spring only to mature and wither with the coming of winter. He had watched the dying of farm animals and he had grieved over the loss of his pet cats and dogs.

Now that his childhood was ending the boy wondered what would happen next. Thinking that this would be the last year of his life, he asked his mother to take his picture as a remembrance of him. Standing on the milk house steps and squinting into the setting sun, he trembled with fear of what the future might hold.

As the sun sank below the far horizon, the boy walked out of the long driveway lined with tall Chinese elms and went down the winding road toward the old place. His ancestors had settled there nearly a hundred years before. They had emigrated from Ireland during the potato famine. On a gentle hill sloping to the marsh they had built their house. They and their descendants had farmed the land.

Before the settlers arrived the land had hosted another people. A tribe of Potawatomi Indians had lived on the oak knoll that rose along the edge of the marsh. Among the trees they had built their houses of saplings and bark. They had fished in the creek that flowed through the marsh and they had hunted in the surrounding woods. The boy's great-grandmother Bridget had told stories of seeing the Indians walking across the field, looking at the place that had once been their home. Eventually the Indians returned no more to the marsh. The boy looked for arrowheads in the plowed fields each spring.

He always had been haunted by the old place down the road. One day when he was very young, he stood beside his father east of the barn and watched as his grandfather, nearing the end of his own life, slowly made his way across the field that sloped down to the old place. The boy was startled by his father's comment in the morning air, "Here comes the old man." An old man was making his way from an old place, this the boy would remember.

There were the subtle reminders that the generations down at the old place had come from Ireland. Bridget, his father had told him, had walked on Sunday mornings to attend mass at the church in town. Great-aunt Kate's glass rosary beads rested in a box of trinkets in the attic of the farmhouse. Although the Irish accent ceased to be spoken, the boy sometimes would hear his father imitate a brogue that was from the original tongue. Faith in things not seen, even of the people of the fairy world, was still in the realm of memory.

The tones and inflections from the old country continued at the house on the hill, at the edge of the marsh, until the lives of the old ones ended. Whenever the boy passed the old place he would hear a sound from another time and another place. He was haunted these early years, as he would be all his life, by the lives of those who had come before him.

Darkness now spread over the marsh and moved up the hill. As the spring night unfolded, frogs at the edge of the pond began to croak. The shrill voice of a nighthawk fell sharply through the sky. A fox barked from the tamaracks in the marsh. Numerous times the boy's father had warned him, "Don't go into the marsh." Finding a passageway under the crumbling foundation of the old house, the boy crawled through and rested in a corner of the cellar into the night.

As the early morning sun rose over the eastern hill and streamed through the cellar window, the boy slowly opened his eyes. He got up and came into the sunlight of the new day. The boy was about to begin a journey that would take him through the marsh in the changing seasons. He would meet strange creatures along the way, and there would be a voice coming from deep within the marsh. The boy exclaimed, "Take me, I am ready."

He crawled through the passageway and walked into the marsh. The mystery was all around him in the singing of the birds, the blowing of the wind through the trees, the sounds of his feet on the earth, and the clouds in the sky. The boy heard the voice telling him to look carefully into the world that he is exploring. "Look through the edge of what is visible and you will see the things that pass you in the dark. You must look twice if you wish to see fully the wonder of the world."

The boy squinted into the morning sun. The light rays shot sharply down onto the surface of the narrow creek that flowed gently through the marsh. For the first time, the boy looked through the edge of the sun's rays and watched as the light danced on the water.

As he looked into the water and studied his own reflection, the voice in the marsh said, "You need certain experiences that will bring you face-to-face with what is true. One begins not by learning something new but by remembering what once was and always will be. In order to know the marsh and all that is in it you must experience the mystery of the marsh."

A ripple formed on the surface of the water, and suddenly water splashed near the bank of the creek. The boy was about to see the creature of trickery and deception. Out of the water and up the soft muddy bank scurried a muskrat. Clutched in his front paw was some of the bottom soil that remained from the creation of the earth. Muskrat snickered and laughed slyly as he advanced toward the boy. "I am not what I seem nor am I otherwise," Muskrat said clearly as he nudged the boy. The boy jumped back another step, tripping over the protruding root of an alder bush. "I am the creature that lives both on land and in the water. I come to confuse you, to show you that things are not always what they appear to be."

For a moment, all the boy had learned to depend on as real during his childhood vanished. Muskrat shook the water from his furry body into the boy's face. Was this an evil act or the natural behavior of an animal? The boy was confused, and he was frightened. He quickly slipped his hands into his pockets to keep them from trembling. The muskrat pushed hard with his hind legs and slipped into the creek and swam away into the reeds and cattails in the marsh.

Each day and night of the passing summer held an adventure. Several times Muskrat reappeared, chuckling and then speaking another riddle. Muskrat continued his attempts to confuse the boy, suggesting the ambiguous nature of everything. Muskrat asked, "Can you hear clearly when you think about listening?" This time the boy answered with his own question, "Can you think while you are speaking?" With this final encounter, Muskrat disappeared into the creek.

The boy was learning to see the truth in the immediate and diverse things of this earth. He was learning to know the marsh. Learning that the meaning of life is to be found not in something far away but in what is very near. The voice within the boy, and within the marsh, was saying, "All that exists is in constant change and is always in the process of becoming something else." After a long silence, the boy said, "I too am changing into something else."

In the late summer afternoon, the boy rested on the hill that sloped to the marsh. He looked into the large white clouds that passed overhead and knew that he was an intimate part of the world.

The days of summer gradually changed their greens and blues into the gold and brown colors of autumn. The sun cast a warm glow over the marsh. Birds gathered in flocks and took flight. Spiders spun their intricate webs among the stalks of the drying sedge. Squirrels scampered from their leafy nests in the oak and hickory trees and gathered acorns and nuts on the ground. Two red-tailed hawks soared overhead on the flow of the wind.

The boy had other things to learn before the leaves left the branches of the trees. Walking deeper into the marsh, he went into the tamarack forest. The voice in the marsh spoke again: "As you grow you will feel a new stirring in your heart. It will come to you as a gift on the wings of beauty."

Suddenly, high on the dead branch of a tall tamarack, Great Blue Heron arched her wings. Ready to fly again a thousand miles into another marsh, she spread her six-foot wingspan and peered down at the boy. He offered her a fish from the nearby stream. Great Blue Heron flew down and landed close to the boy and accepted the fish.

Her soft blue feathers brushed against the boy's legs. He reached out and stroked her long graceful neck. From her throat she whispered softly, calling as she did to those she loved. He felt a stirring in his heart of what it might be like to love one as beautiful as Great Blue Heron. She whispered to him, "I will return to you another day." She raised herself and, throwing back her neck, opened her wings and flew into the sky. For a moment the boy felt himself flying with her, suspended between heaven and earth.

He suddenly remembered his mother and father back on the farm. For an instant he felt their affection pulling him back toward childhood. A voice sounding like his mother's cried, "Please stay with me." The boy began to run, first in a direction that led to the farm, then in another that took him back into the marsh.

The late afternoon wind from the west was blowing hard as he came out of the tamaracks. The harsh wind chilled the boy's face. Winter was coming. The boy made a shelter for himself on the southern slope of the oak knoll.

The snow fell quietly, laying its soft white coat over the marsh. As the sun climbed above the hill to the east black crows in the top of the hickory tree called out. A winter mole burrowed deeper into the snow. Willows laced the lowland. High in the sky wild geese honked in formation as they passed over the marsh. The boy stepped down to the water's edge and washed his face in the icy water. He stood erect, facing the morning sun as it arched in the sky.

The boy left the shelter and explored the new winter day. Walking up the hill among the oaks, he came to the mounds where Indians had been buried.

Bright stars shimmered above in the stillness of the cold winter night. A full white moon cast a light on the hard surface of the snow. Cattails stood silhouetted in the frozen marsh in the still of the night. The voice in the marsh spoke once again: "Silence is the dry leaf that does not stir in the wind. The mystery of the world is near to us when we are still." The boy was learning to keep his mind quiet and calm.

As morning neared a dense hoary fog settled over the marsh. Shortly a wind began to blow out of the north. On the hill the frigid wind moved through a gully, long ago washed by the rains and melting snow. Suddenly, over the horizon, swirling in the wind, appeared the Winter Witch. She moved in beauty but chilled the body and the soul. In a lacy white gown covered with crystals of ice, Winter Witch circled the marsh and settled down on the other side of the gully. With her shining blue eyes, she stared at the boy, chilling him to the bone. He pulled his cap farther down on his head and zipped up his jacket.

Winter Witch spoke. "There are new lands to be won and souls to be taken." And then, her voice crackling like ice and thunder, she screeched, "I have come to take you." The boy felt her breath pass through him. A barren soul in lace and crystal, she stood very near to him. "I have conquered death in my own way," she said.

"You can have everlasting life, your own self forever, in your present form, if you come with me," said Winter Witch, her voice rising to the pitch of a scream. "You can be free of the ways of nature if you follow me. Winter can be yours forever." She reached for the boy's arm and pulled him toward her.

The boy and the Winter Witch struggled throughout the gray dawning hours. Falling to the ground they rolled down the snow-covered hill. When daylight came to the edge of the marsh, the boy lay locked in the icy arms of Winter Witch.

Signaling its place in nature, the sun began to rise over the eastern hill. Winter's first warm breeze came out of the northwest sky. The boy slowly regained his strength, and with a strong thrust of his arms, he pulled himself free of Winter Witch. He shouted into the air, "I do not want to stay the same. I will not go with you." The boy watched as the crystalline gown of Winter Witch began to melt in the warm breeze that filled the marsh.

On the last cool current to pass through the gully on the hill, Winter Witch vanished from the marsh. "A dead soul you must guard against all your life," the voice of the marsh said to the boy as Winter Witch rose into the few gray clouds that remained in the warming sky. "The only everlasting self is the one that is in harmony with nature. The true self is always growing and changing, coming finally to be a part of all that is."

How strange this seemed to the boy the first time he heard it. He thought to himself, "I will never be taken by the Winter Witch." The last snow of winter was melting on the hills and flowed slowly into the marsh.

The warmth of new life spread over the land as the spring sun reached halfway into the sky. Over the far horizon to the south, Great Blue Heron appeared, her wings embracing the warm current of the marsh air. The slopes were turning green with fresh and tender grass. In the woods the shooting stars raised their heads toward the sun. The mandrake plant spread its leaves over the early-forming fruit. On the hill north of the marsh the soft leaves of the mullein pushed through the earth.

Perched on the new cattails in the marsh, red-winged blackbirds sang their songs.

In his journey throughout the year in the marsh, the boy had grown to love the marsh and all its creatures and living things. He learned that only through birth, growth, and death does life continue endlessly. Change is a necessary part of the great mystery of life on earth. In the years of travel that would follow, what the boy had learned here would serve him well. "I feel at one with the mystery of the marsh," he said. "I am changing and I am becoming a part of the wonder that is all around."

The last hours of the boy's adventure in the marsh were drawing to a close. He would be missed on the farm if he did not return soon. He walked slowly toward the old place in the orchard on the hill. When the boy reached the foundation of the old house, he climbed through the passageway where he had entered so long ago. On the other side the farm was in sight. With a pace more certain than when he had gone to the old place, the boy walked toward the farm, gazing with anticipation all the time at the unexplored horizon that spread beyond. The sky glowed as he quickened his pace for the return home. He knew his parents would be waiting for his safe return.

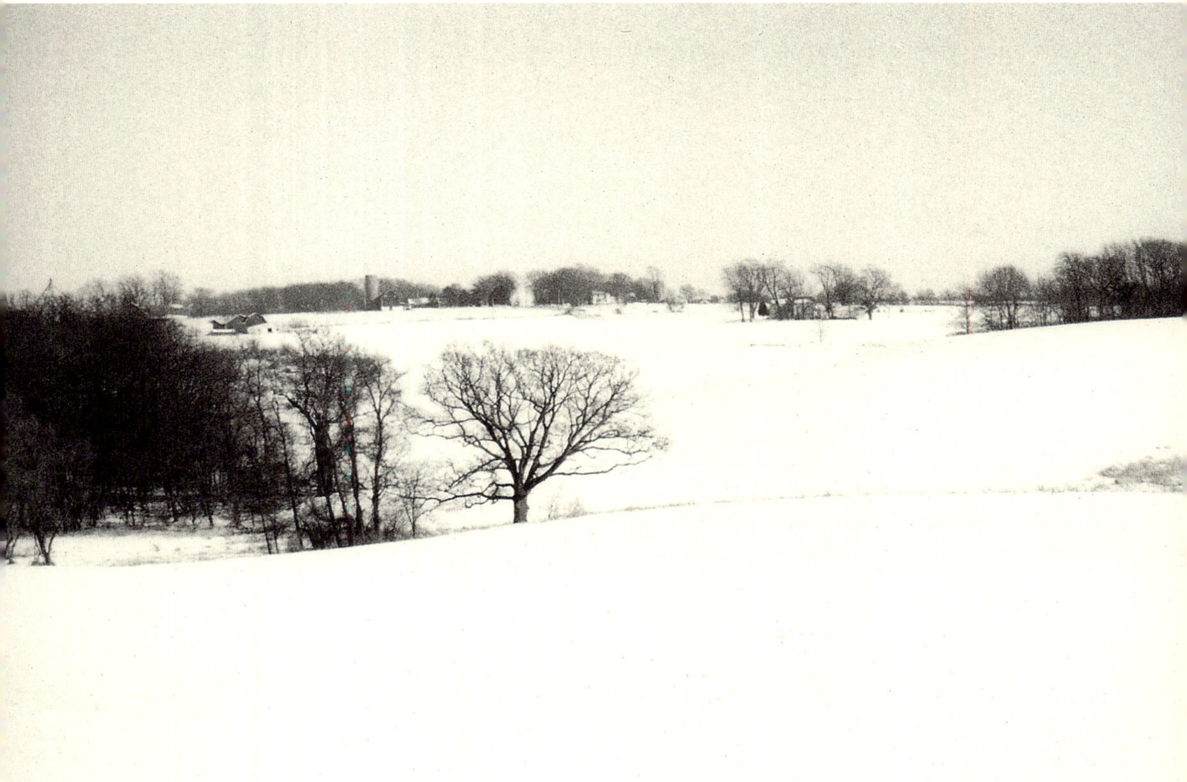

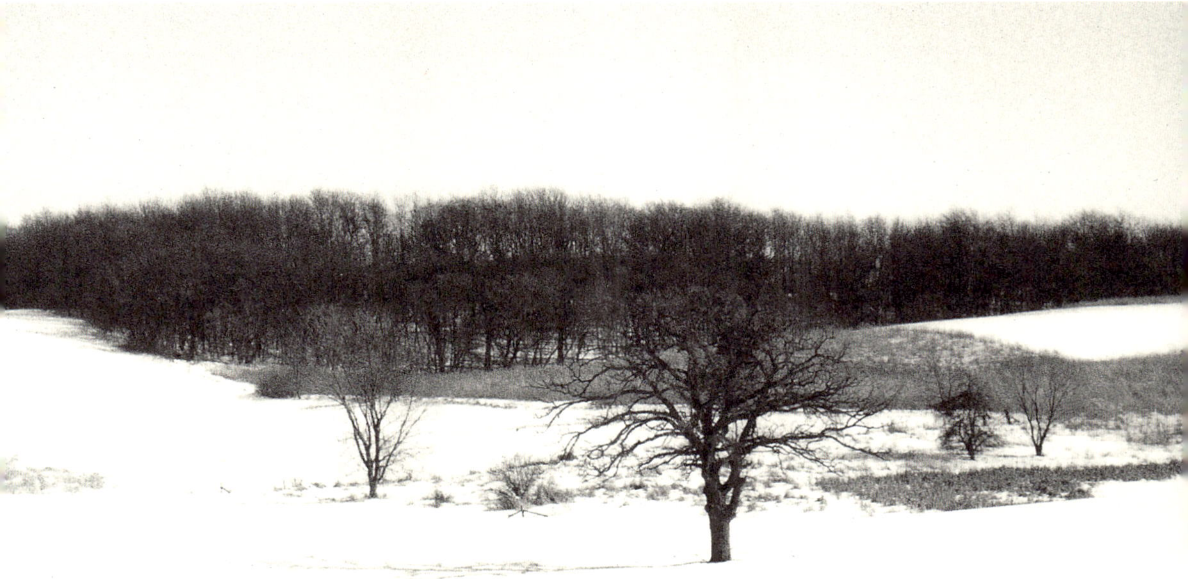

Part I

The Nature of All Things

> Try to be mindful, and let things take their natural course. Then your mind will become still in any surroundings, like a clear forest pool. All kinds of wonderful, rare animals will come to drink at the pool, and you will clearly see the nature of all things. You will see many strange and wonderful things come and go, but you will be still. This is the happiness of the Buddha.
>
> Achaan Chah, *A Still Forest Pool*

A soft morning light. In the distance, the ancient bur oak, silhouetted and solitary in the middle of the pasture, casts a shadow on the ground. The family of four sandhill cranes, parents and this season's young, graze in the grass beside the pond. The old place where my great-grandparents settled nearly a hundred and fifty years ago rises at pasture's edge. Pioneers, then, as we hope we are now, still this late in life.

Sometimes, in fleeting moments not usually willed, you will grasp somewhat the wonder and essence of nature. You will then sense that there is no separation between nature and us.

That we are nature, that all things in this ever-changing world are of nature's domain.

These field notes are made in the spirit and thought of Ralph Waldo Emerson. In the break from the traditional doctrines of religion, as Harold Bloom has noted, assuming the crucial components of Emerson's American religion: solitude, God within, and the oldest part of the self going back before creation. For Emerson and the poets that followed, Emily Dickinson and Walt Whitman, there is the uncertain border between the living and the dead, between the animate and the inanimate. In the woods, Emerson wrote, "Standing on bare ground—my head bathed by the blithe air, and uplifted in infinite space—all mean egotism vanishes. I become a transparent eyeball; I am nothing; I see all; the currents of the Universal Being circulate through me; I am part or particle of God."

I have been informed and inspired throughout my life by the life and writing of Henry David Thoreau. Nearing the end of his life, he said that he was content living in this one world. As much as we might look to the heavens, this earth is the only world in which we move and have our being. Where we might see into the nature of all things.

We remind ourselves all the while that nature is not necessarily benevolent to our human interests. We are subject to the same forces that work in all of nature. Nature is the source of our life as well as its destruction. As new life is being created. With pen in hand and camera at my side, traveling between town and country, I make my notes and live this day. One world at a time, here on earth.

The words are repeated in prayer every day by millions of people. Beginning, "Our father which art in heaven," and ending "For thine is the kingdom and the power and the glory for ever." The prayer, treasured from the time of early Christianity, Jewish in form and content, is addressed to an omnipotent God in the practice of a holy life. Individual interpretations and intentions vary in a multitude of ways. For many people, little or no attempt is made to connect to a deity that even metaphorically can be identified as a personified being beyond this world. It is enough to know that we are an integral and intimate part of a great mystery.

In the middle of the sermon that the prophet Jesus offered on the Mount, we find our spiritual center: "On earth as it is in heaven." The kingdom of heaven, as much as we will ever know of it, is created and lived here on earth. This search, this travel, this life of making a heaven on earth, is the life of the pilgrim.

The *Tao Te Ching*, the classic Chinese manual on the art of living, informs the pilgrim that a long time ago, before heaven and earth, there was a great mystery. There was something formless and complete. Solitary and unchanging, all-pervasive, infinite, and without name. Only in the naming, in the naming that we humans give to things, including the mystery, do distinctions and judgments arise. The mind is mother to the ten thousand things.

Lao-tzu, the attributed author of the *Tao Te Ching*, gave us the model of an effortless and spiritual life. A life of pure and effective action sometimes referred to as "non-action."

> Less and less do you need to force things,
> Until finally you arrive at non-action.
> When nothing is done,
> Nothing is left undone.

When we trust the world, when we have faith, there is nothing that has to be done, and yet everything gets done. The pilgrim, in harmony with the way things are, has no need to search. And there are many discoveries and adventures to be found along the way.

Whatever you may see and whatever you may write is a witnessing to natural history. Whether you are the country clergyman Gilbert White recording the daily lives of birds and attending to local natural occurrences in Selborne at the end of the eighteenth century, or you are Henry David Thoreau observing the seasons from a cabin at the edge of Walden Pond in the nineteenth century, or you are the Buddhist monk Achaan Chah in a forest monastery in Thailand meditating, talking, and writing about being mindful, or whether you are carefully living and observing each day, and perhaps writing, you are participating in and contributing to natural history. When you are in the world of things as they are, you are in the world of natural history.

As you sit in the doorway of your cabin, or as you stand at the corner waiting for the light to turn, you will experience an awareness of this present and wondrous moment. The attention you give to what is here and now will give you all the transcendence you will ever need. At the end of *Walden,* Thoreau exclaims: "Only that day dawns to which we are awake. There is more day to dawn. The sun is but a morning star." You know that you are part of a vast and unknowable universe. A world without name, before heaven and earth, here and now, without end.

Savannah sparrows, red-winged blackbirds, and kingbirds are nesting in the field this summer east of the barn. I attend to the occurrences in this kingdom of which I am an intimate part.

A line from the *Meditations* of Marcus Aurelius: "Oh Nature! From thee all things, in thee all things subsist, and to thee all tend." All things that we know, and yet do not know, are of nature. Our own human being included, intimately and ultimately. There is no separation between nature and us. Only misperception makes nature something other than the reality.

The etymology of the word *nature* begins with the Latin *nasci,* "to be born, to spring from, to arise, to be produced." Nature, then, as Donald Crosby reminds us in his treatise *A Religion of Nature,* "is the creative matrix from which all things arise and to which they return, the complexity of orders and powers by which these things are upheld and by which each of them, or each type of them, attains its own peculiar attributes and capabilities." There is nothing outside of nature, beyond nature, or before it. Everything that exists, that has existed, that will exist is an integral part of nature.

Whether we humans want to make a religion out of nature depends upon our individual and collective need to have a religion of any kind. As I walk these summer fields, and wander into the woods at the edge of the moraine, lost in the moment and in the beauty of this world, there seems to be no need for any other religion.

Only a farm away, to the north, is the newly created Turtle Valley Wildlife Area. The restored marshland that was drained for farming in the early 1900s has become home to great numbers of bobolinks, upland sandpipers, and other grassland birds, as well as blue-winged teal and a host of other waterfowl and shorebirds. Habitat has been improved for threatened and endangered species, including American bittern, Wilson's phalarope, king rail, Blanding's turtle, and eastern massasauga rattlesnake. This spring three migrating whooping cranes rested in the marsh for several days before continuing their flight to central Wisconsin. A wetland of twenty-three hundred acres is being preserved for future generations of plants and animals, including us humans.

We stand on a great watershed. Turtle Creek flows west and south to the Rock River, and Sugar Creek flows to the Fox River and into the Illinois River. Both streams that originate in Turtle Valley flow into the Mississippi. The rivers then flow into the Gulf of Mexico, becoming part of all the waters of the world.

We are the latest to care for this piece of land. On a land whose recent geological history starts with the beginning of the Ice Age about two and a half million years ago. The last series of glacial advances began about twenty-five thousand years ago, and ended about ten thousand years ago. We are in the midst of the terminal moraines of the last glaciers. The moraine ridges a mile to the north, part of the Kettle Moraine, were formed from the materials deposited as the glaciers receded with the warming of the climate.

Then came the new plants and animals, followed by the various periods of human occupation: Paleo Indian, 10,500 to 7500 BC; Archaic Indian, 7500 to 500 BC; Woodland Indian, 500 BC to 1600 AD; Historic Native American Indian, 1600 AD to the present; and Settler Occupation, early 1800s to the present.

Emigrants from European countries settled here in Walworth County from the 1820s onward. Settlers came from the eastern states of New York, Maine, Vermont, and Virginia. Their countries of origin were predominately Norway, Germany, England, and Ireland. To the south, on the hill, is the Sugar Creek Lutheran Church, still the center of the Norwegian community.
To the east is the site where my great-grandparents James and Mary Holloway lived for a time in a house at the edge of this

marsh—at the end of Connaughton Road—after emigrating from England in the 1860s. Even at this time, marsh grass was being harvested in the valley. Plowing of the marsh for other crops would begin with the new century.

The vast wetland prairie would be drained for farming with an extensive network of ditches and drain tile. Because the soil, primarily Houghton Mulch, had high amounts of organic matter, it was very fertile and produced good crop yields. In dry weather, the soil dried to a powdery fluff that eroded with the wind. Heavy black clouds of dust would darken the skies and fill houses nearby with black dust. And peat fires would burn underground for weeks at a time.

Labor-intensive farming began in the valley in the mid-1930s. Potatoes, carrots, and onions were grown and harvested in an operation that spanned much of the valley. Workers from Mexico and Texas came to work, beginning as seasonal laborers, and then some of the families purchased and operated the farm. When the history of the Hispanic population in Walworth County is written, Turtle Valley will be a primary source of that history. Farming of the marsh ended with the growing and harvesting of mint by still other settlers. A new era in the valley was about to begin.

The marshland is now being transformed, becoming something like it was thousands of years ago before being occupied by settlers. The goal is simple and elegant, the creation of a harmonious balance between plants, animals, and people. A balance between the wild and the domestic, a harmony between utility and beauty. In the restoration of Turtle Valley is the recognition that people can be healthy only when the landscape is healthy. We now share the valley with all the other species of plants and animals.

The relation of humans to nature is noted succinctly in Crosby's *A Religion of Nature:*

> Human beings are an integral part of nature, one biological species among an extraordinary profusion of such species across the face of the earth. Every species has something distinctive and special about it; that is, of course, what warrants its identification as a species. Some species, however, are more distinctive than others, and none is more astoundingly different from other kinds of organisms in its behavior, history, and accomplishments than the human species.

A distinctive and amazing species, certainly, but like all other species, the human species is bodily bounded, embodied in nature. Breathing, eating, experiencing suffering and pain, sleeping, reproducing, dying—as with species of all kinds. Bodily existence is our intimate link to other living creatures on earth.

Creatures, animal and otherwise, are part of the inorganic world as well as the organic. Alfred North Whitehead considers the single molecule: "It is part of nature. It has moved about for millions of years. Perhaps it started from a distant nebula. It enters the body; it may be a fact in some edible vegetable; or it passes into the lungs as part of the air." We humans, in such realization, know that we are of the stars. We began life in some faraway place, a long time ago.

Humans share with many other species the DNA material that controls the genetic makeup of each species. We share with the chimpanzee, our closest relative, 98.4 percent of our nuclear DNA. The genetic connection extends to the plant and vegetable species with the humbling report that 75 percent of the genetic makeup of humans is shared with the pumpkin. The point being that humans are more of nature than conventionally thought, that we are indeed nature itself. There is no separation between human beings and nature. We are nature.

We humans share, with other creatures of nature, social behavior, culture, and consciousness. That these characteristics of life are common outside the human species has been ignored and unknown until recent times. As if ignoring the social life, the culture, and the consciousness of other species would confirm the distinctiveness of humans in comparison to other beings. Yet, even the trees and flowers, we now know, have a consciousness as leaves and blossoms direct themselves to the sun as it moves across the sky each day. Consciousness, social behavior, culture—only a difference of kind and degree between species.

Still, there is a remarkable difference in degree between the human species and other species. The ability of human beings to reason and speculate has given rise to unique cultures and civilizations. With a complex language, humans have become elaborate storytellers and creators of elaborate myths by which we live our lives. We humans are constantly changing our environment, and may be setting in motion forces that will endanger the environment for all species. May the ethic of diversity, balance, and health prevail.

This summer's nesting of red-tailed hawks send haunting screams of *keeyah* from top of the old oak in the woods at the edge of the pasture. In this part of the county an attempt is being made to create and maintain a healthy habitat for all creatures.

A memorial service is held for Gaylord Nelson in the rotunda of the state Capitol. The service began at noon and drew more than four thousand people. U.S. congressmen, former colleagues and friends of Gaylord Nelson, his daughter Tia, and former Vice President Walter Mondale gave eulogies. Representative David Obey brought the audience to loud applause when he declared, "If the Senate had voted the way Gaylord did, there would be one less memorial on the National Mall." He was referring to the Vietnam War, but clearly implying the consequences of the on-going war in Iraq.

Nelson served his last years as counsel for the Wilderness Society. He had been Wisconsin's governor from 1959 to 1963, and served three terms as U.S. Senator from 1963 to 1981. But it is as a founder of Earth Day that he is best remembered. And the Apostle Islands off the shores of Lake Superior have been preserved as a national park largely through his efforts. The earthly remains—the ashes—of this revered man are to be placed in the family plot in Clear Lake.

The printed program for the memorial service quoted two passages from Nelson's first Earth Day speech in April 1970:

> Our goal is an environment of decency, quality, and mutual respect for all human beings and all other living creatures—an environment without ugliness, without ghettos, without poverty, without discrimination, without hunger, and without war. Our goal is a decent environment in its deepest and broadest sense.
>
> The battle to restore a proper relationship between man and his environment, between humans and other living creatures, will require a long, sustained, political, moral, ethical, and financial commitment far beyond any commitment ever made by any society in history. Are we able? Yes. Are we willing? That's the unanswered question.

The service ended with a haunting harmonica rendition of "Amazing Grace."

The Fjord horse grazes in the pasture south of the house. Her ancestors migrated to Norway from Western Europe over four thousand years ago. The breed served farmers for centuries along the fjords of Norway. Archaeological excavations contain the remains of these horses. Gunnel, as she is named on the registration papers, has the genes of the wild horses that once roamed the mountains and plains of Asia, long before the Vikings, and long before this fenced pasture south of the farmhouse.

This field has been without horses for sixty years. Workhorses of mixed breed had been tended by the three generations of my Irish ancestors. The love of horses flowed in Irish blood. Horses were bought, bred, trained, and worked on this land. Rusted horseshoes are still pulled from the ground.

The Catholic faith of my great-grandparents has been supplanted by an intellectual and spiritual tendency shared by a great many others in the years since the famine and emigration. A temperament that is religious without an absolute religious belief. But a religious temperament nevertheless—a temperament that is deeply spiritual. Nature as the source of all being.

Belle, the fourteen-year-old white mule, is now companion to the Fjord horse Gunnel in the newly fenced pasture. Belle gently nudges Gunnel as the two graze about the field. To keep late August flies from biting her back, Gunnel rolls in the grass. Apples from the old trees in the orchard furnish a treat offered from outstretched hands.

There are new signs of life on the farm. A sense that the old era has passed and that something new is beginning. After the team of workhorses was retired sixty years ago, sent to the mink farm for slaughter, the farm was mechanized with the latest in power machinery. Eventually the farmer and his wife retired from farm work, and the land was rented for cash crops. The sons moved to other places and to other lives and livelihoods. Now something new is rising from the ruins of this land and this old place.

We could be characters in a play by Anton Chekhov. We would be on an estate in Russia, a century ago and a continent away. Astrov, the doctor who regularly visits the estate, is talking about the destruction of the forests. He laments the shortsightedness of his fellow inhabitants of the land. He speaks to the family gathering: "Man is endowed with the reason and creativity to increase what's been given him. So far all he's done is destroy it. There are fewer and fewer forests, rivers are drying up, game animals are all but extinct, the climate's being ruined, and day by day the land gets poorer and uglier." Regarding his own efforts, Astrov adds, "If man is happy a thousand years from now I'll have played my little part."

Who would not hope for an improvement in the land, and want to play some part in making a better world? A world that is healthy for all plants and animals. This is an ethic that can be practiced and demonstrated on the farm that spreads around us in all the directions of earth and heaven.

Snow—all night and all day—falls generally over town and country. I remain cloistered all day in my house in town, becoming accustomed to weather and season. Trees barren of leaves line the city streets. Days once cast in golden light have given way to the gray shades of winter. Nature, inside of us and out, takes another turn.

There is nothing like the new snowfall to enter a space that is quiet and seemingly empty. The world appears, for stretches of time at least, to be beyond our comprehension. We realize, as John Fowles wrote, in his long essay "The Tree," that nature is truly alien: "It may sound paradoxical, but we shall not cease to be alienated—by our knowledge, by our greed, by our vanity—from nature until we grant it its unconscious alienation from us." The mythic "green man" of the tree, the wild nature within us, is realized, beyond the naming of the wonders of the world. Only as we see ourselves as part of nature do we begin to find our way in the natural world.

This giving of close attention to a limited and circumscribed geography is in the tradition of the country cleric who watches over the parish. At the end of the eighteenth century, Gilbert White published the results of his years of close observations in an English village in *The Natural History of Selborne.* Confining his observations to one small area, White, as his biographer Richard Mabey writes, "showed how watching the natural world at close quarters could generate not just understanding but respect, and an insight into the kindredness of living things." White's influence has been far-reaching. As his biographer notes, his work "helped foster the growth of ecology, and the realization that humans were also part of the natural scheme of things." White has given inspiration to generations of nature writers and ecologists.

Let this area of southeastern Wisconsin, from town to farm, be the site for observations large and small. Seasons will come and seasons will go. All things observed, usual and unusual, will be seen as part of the natural scheme of things. This will be, as Gilbert White wrote, a "parochial history." The oak now stands in a first layer of snow, a few brittle brown leaves moving in the wind.

The search for meaning is as transitory as what is being searched for. The more one looks, the less there is to be seen. "Reach, and it can't be grasped," cautions the *Tao Te Ching*. But know where you came from.

> You can't know it, but you can be it,
> At ease in your own life.
> Just realize where you came from:
> This is the essence of wisdom.

This is the place where wooded groves are scattered among glaciated hills and prairies. This is the land that Donald Culross Peattie eloquently describes in *A Natural History of Trees*. In the section on the bur oak, he tells of the pioneers who came to the Middle West and found a new type of forest growth. They viewed for the first time groves of wide-spaced trees over carpeted fields of grass. And dominant in these groves, in these wooded openings, was the bur oak.

> The most characteristic tree which the settlers noted in the Oak openings was one which grows, indeed, to the eastward, but never comes there to its full grandeur or its characteristic park-like vistas. They know it as the Bur Oak because the heavily fringed cups of the acorns look almost like the burs on a Chestnut, though the cups only half cover the nut. True that other Oaks are often found in the Oak openings, or prairie groves, of the Middle West, but ecologists consider them secondary successors; they have come in after the Bur Oak has conquered the prairie for them.

The bur oak is the pioneer among the trees that advanced over hills and grasslands. With its extensive root system, keeping other trees at bay, the bur oak often stands alone in the field. Dark against the horizon, stout trunk, branches spreading and crooked, rounded crown, this is the tree that is firmly grounded in native soil, the tree that is deeply rooted in the heart and spirit of any pioneer who has spent time here.

It is the human condition to desire what is not yet known. A predilection honed by evolution to imagine things beyond present experience. Scientists entertain the possibility of other dimensions, and posit other universes beyond or within our own. Theologians, and the practitioners of many religions, believe in a kingdom beyond earthly existence. Whether by reason or faith, we humans want more than we have, envision more than can be seen. A day at the farm prompts the desire to return to the city. Back in town, one looks forward to other days in the country.

No matter how much Richard Jefferies wrote about nature in his native England during the latter part of the nineteenth century, mysticism took him to realms beyond the ordinary world. Nature was seen as a mystery, and Jefferies imagined a world beyond the gods. In his book *The Story of My Heart,* published in 1883, he wrote of becoming part of a higher realm, entered through the natural world: "I now became lost, and absorbed into the being or existence of the universe. I felt down deep into the earth under, and high above into the sky, and farther still to the sun and stars. Still farther beyond the stars into the hollow space, and losing thus my separateness of being came to seem like a part of the whole." Jefferies desired a "soul-life," a strengthening of the emotion that bound him to the wholeness of the world.

Wherever his body and soul may be after death, Jefferies enjoyed the thought of his own immortality. "If then, after death, I am resolved without exception into earth, air, and water, and the spirit goes out like a flame, still I should have had the glory of that thought." Jefferies was sick and dying of consumption as he wrote the biography of his soul, and did not have long to live.

True to his vision of nature, Jefferies made his last request:

> Could I have my own way after death I would be burned on a pyre of pine-wood, open to the air, and placed on the summit of the hills. Then let my ashes be scattered abroad—not collected in an urn—freely sown wide and broadcast. That is the natural interment of man—of man whose Thought at least has been among the immortals; interment in the elements.

He thought that there is something more than existence. "There is so much beyond all that has ever been imagined," he wrote. What we know as nature is already invested with the supernatural. Ultimately nature cares not for the human being. Jefferies wrote: "As these natural things have no connection with man, it follows that the natural is the strange and mysterious, and the supernatural the natural." He ceased to look for traces of a deity, finding no traces in the world of nature. Jefferies concluded that there must be something even higher.

And what does your narrator of heaven on earth have to say about this? Simply, that there is enough to be known, with the gift and power of human imagination, to have all the mystery one needs in the living of this daily life. The extraordinary, with imagination, is found in the ordinary world.

No need, no reason, to look beyond this world where we now live. Henry Thoreau reminded us of this in the last days of his life. It is reported, as his biographer Robert

Richardson Jr. informs us, that as Thoreau lay dying in his bed, a family friend and former minister by the name of Parker Pillsbury gently asked him the question, "You seem so near the brink of the dark river that I almost wonder how the opposite shore may appear to you." Thoreau answered, summing up the philosophy and the living of his whole life, "One world at a time." Might we who move between town and country, in this life, repeat and practice the words that Thoreau uttered from his bed.

The story of Seymour Reynolds is still told in these parts. He was a bachelor brother who had moved into the small house by the lilac bushes after Hattie, the wife of my grandfather John Quinney, had died, leaving two young children and a baby.

We are told that Seymour loved horses and raised colts. He liked to play cards, and in the winter of 1913, he and some neighbors played late into the night. After the neighbors had gone, he found that there was no wood in the house to keep the fire going. He went to the woods, and on his return to the house, walking over the snow-covered field, he slipped on a patch of ice and fell into the snow. As he fell, he was struck under the chin by the log he was carrying. He suffered a broken jaw and lost consciousness from either the fall or the freezing temperature in the night. The dog that accompanied him lay faithfully beside the body for three days until the discovery was made.

That is all that anyone today remembers about the man, Seymour, who once crossed this field on a cold winter night. The great bur oak that he would have passed still stands midway in the field.

Wallace Stegner has written this: "So I must believe that, at least to human perception, a place is not a place until people have been born in it, have grown up in it, lived in it, known it, died in it — have both experienced and shaped it, as individuals, families, neighborhoods, and communities, over more than one generation." And the ancestors have lived, as real people, only as long as they are remembered. Only as long as the stories are told by those who remain.

Walking down the hill through the snow-covered field to the bur oak — I am a small figure against the landscape. The scene could be from a Chinese painting of the Ming dynasty. For a moment I sense what Albert Camus alluded to as the indifference of the universe. Someday my life as a conscious being will end and I will become part of the universe that is indifferent to my fate as a singular human being.

Against this winter landscape, and without some other consoling thought, we are humble servants to natural forces beyond our comprehension. Behold, this sacred space, outside the knowing mind. For an instant, an instant only, I have a sense of my place in the vast and unknowable universe. A chill runs through mind and body — fear, awe, and wonder. Ultimately, you must surrender.

The nest in the oak is empty. Perched high on the wire beside the road, the wintering red-tailed hawk waits for the mole to scurry across the snow. One swift swoop to the ground and the hawk has satisfied a hunger.

Maybe it is the weather, or maybe it is the aging, that draws us to a realm beyond the human. Beyond this human self that we have come to know, and to love, so well. As much as we value the human and the humane, a natural world much larger than human existence now opens. To be human, truly human, is to realize our part in all that is, nonhuman as well as human. Our humble human being in a cosmos in which we have our beginning, our ending, and our eternity.

For inspiration, and consolation, I return to the poetry of Robinson Jeffers. He wrote poems of austere beauty during a life on a rocky cliff above Carmel Bay. He found freedom in the notion that human life and human suffering have no special meaning in the larger scheme of things. In an early narrative poem titled "Roan Stallion," published in 1925, Jeffers wrote:

> I say
> Humanity is the mold to break away from, the crust to break
> through, the coal to break into fire,
> The atom to be split.

We humans are to sense the violence at the center of existence. From the hawk's claw to the fury of war to the breaking of stones. Finally there is a leap out of the human to the wholeness of things, and to an understanding of the indifference of nature. "Inhumanism" is the name Jeffers gave to this revelation. His verses, he said, "express a protest against human narcissism."

A vision as cool and aloof as a hawk. Still, we remain haunted by the riddles of desire and suffering, by the meaning of our unknowable lives. Relieved only by a sense of the oneness of the natural world. There is spirituality in this that survives the death of theologies. A terrible and satisfying wonder as the hawk returns to the tree in the snowy field.

A leap of faith, it is, to accept fully the prayer William Wordsworth offered to his sister Dorothy in lines composed a few miles above Tintern Abbey:

My dear, dear Sister! and this prayer I make,
Knowing that Nature never did betray
The heart that loved her.

Here in town snow swirls around the pine on its way to the ground already deep in snow. The birdbath is covered now in snow. A car occasionally comes up the hill, making a path out of this dead-end street. Inside, one stays warm, and moves about in little more than reverie. Perhaps a few words will be made to the day. Why even this? On writing, in a book by that name, R. V. Cassill makes a guess: "Writing is a way of coming to terms with the world and with oneself. The whole spirit of writing is to overcome narrowness and fear by giving order, measure, and significance to the flux of experience constantly dinning into our lives." Yes, the flux, and the flotsam and the jetsam.

Is it only a dream? This life that we think we know, think we know with a restless mind. This life, this planetary life, may be but an illusion. Still a life most wondrous. Neil Young sings his song:

It's only a dream
And it's fading now
Fading away
It's a dream
It's only a dream
Just a memory
Without anywhere to stay.

Dreams fade away. We wait for the next one to come. One

illusion of what it's all about giving way to another. Death is the ending of a long line of dreams.

How would Samuel Palmer, in mid-nineteenth century, paint or etch the great bur oak? He would be looking at the tree as much with his inner eye as outer eye. The image of the tree in the field would have been inspired by the visionary poet and artist William Blake, friend of Palmer. On the canvas, or in the etching, we would be witnesses to visual boldness and a poetic vision.

We would perceive the animate form within nature. With pen, ink, and watercolor, inanimate nature—muscular, twisted, and furrowed. The inner light of nature is made visible to the human eye. I approach this tree, this oak, in a contemplative mood. And meditate on the intricate beauty and mystery in this world. Standing in awe once again before the great bur oak.

Here on the farm, or here in town, we contain multitudes. We are sometimes in the sublime of Walt Whitman, inhabiting a world that is wildly diverse. As Whitman writes in a portion of *Leaves of Grass:*

> My faith is the greatest of faiths and the least of faiths,
> Enclosing worship ancient and modern and all between ancient
> and modern,
> Believing I shall come again upon the earth after five thousand
> years,
> Waiting responses from oracles, honoring the gods, saluting the
> sun,
> Making a fetish of the first rock or stump, powwowing with
> sticks in the circle of obis,

Helping the lama or Brahmin as he trims the lamps of the idols,
Dancing yet through the streets in a phallic procession, rapt and austere in the woods a gymnosophist,
Drinking mead from the skull-cup, to Shastas and Vedas admirant, minding the Koran,
Walking the teokallis, spotted with gore from the stone and knife, beating the serpent-skin drum,
Accepting the Gospels, accepting him that was crucified, knowing assuredly that he is divine,
To the mass knelling or the puritan's prayer rising, or sitting patiently in a pew,
Ranting and frothing in my insane crisis, or waiting dead-like till my spirit arouses me;
Looking forth on pavement and land, or outside of pavement and land,
Belonging to the winders of the circuit of circuits.

Leigh Eric Schmidt tells us, in his book *Restless Souls*, that in these lines of poetry we "move from ancient paganism through indigenous traditions of North America and Africa through Buddhism, Hinduism, Christianity, Islam, and Spiritualism." We all are invited to Whitman's "Religious Democracy," as we travel this open road. And we, Whitman informs us, are to find our own way: "Not I, not any one else can travel that road for you, / You must travel it for yourself."

We walked to the barn, flashlight in hand, to see if we could find Gunnel and Belle as the new year was to begin. With the beam of light pointing into the dark field, four eyes shone back into our eyes. Horse and mule galloped toward us and soon were in the barn receiving from our hands the treats we had brought to them. The old barn had not known such life for half a century.

The next morning, with faint light in the gray sky, I photographed the bur oak and the remains of the old place at the edge of the field. Solveig, with sharp eyes, spotted a form in a distant tree. Walking through the snow-covered field, we reached the tree and found an opossum resting in the branches, carefully watching us as we watched it. We followed the road back to the house as Gunnel and Belle dug in the snow in search of last summer's crop of hay. We filled their barnyard feeder with hay and kernels of corn. And made certain thawed water was available in the tank beside the old milk house.

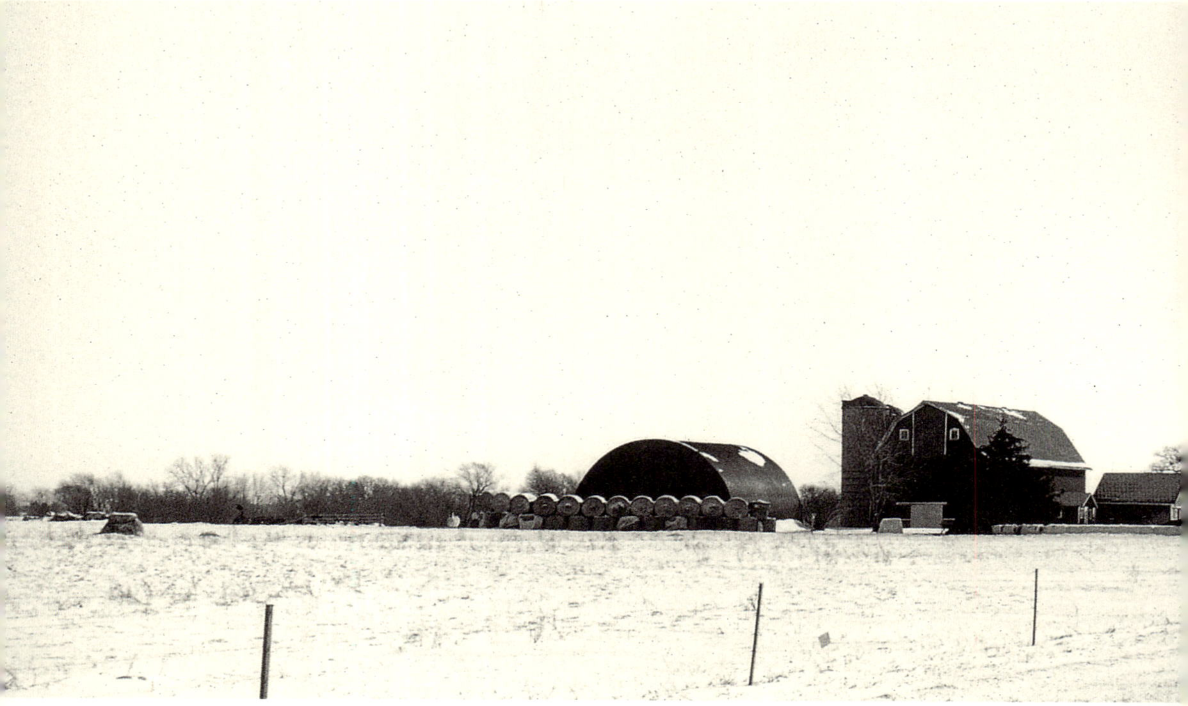

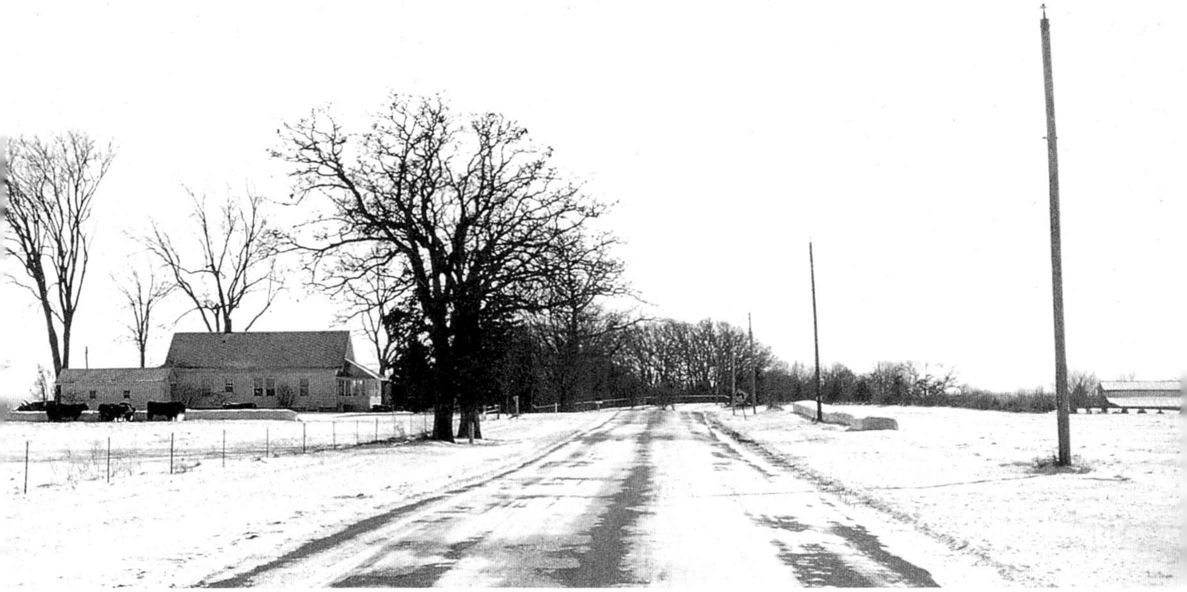

Part II

A Work of Art

> Be alert to the senses. Elevate the ordinary. Art is about a heightened state of awareness. Try to treat everyday life, or at least parts of it, as you would a work of art.
>
> Michael Kimmelman, *The Accidental Masterpiece*

Not so long ago, I was cutting grapevines from the branches of the oak tree at the bend of the road when my leg went limp. A neighbor passing by saw me pulling myself along the fence line and brought me back to the farmhouse. I was having a brain hemorrhage, thirty-five years to the day since my father had a stroke and died while working on the farm. After an emergency call, and a brief wait, I was taken out the front door of the house and carried across the lawn to the waiting ambulance, just as my mother had been carried the day she died five years before. This is a cautionary tale, but what caution is to be taken? More of a reminder of the preciousness, and precariousness, of life in this one world.

Nowadays I tire more easily than I would have a year ago. A trek across the thirty-acre field leaves me exhausted, and I must wait awhile before beginning the return up the hill. Home again, I rest in the house for at least an hour before resuming work of any kind.

The same fatigue occurs as I—on another day—walk over the cover of snow on Monona Bay to watch and to photograph men fishing through the ice. The state's Capitol rises majestically in the distance along a horizon of buildings old and new. Men in grave silence sit on buckets, fishlines dangling into dark watery holes cut through the ice earlier in the day. Warmly dressed—in colors red, yellow, blue, orange, and camouflage—the fishermen wait patiently on the surface of snow and ice.

Why this ice fishing, and photographing? Simply, beyond any other reason, fishermen fish and I photograph to live intensely. Through our work and our play we elevate our daily lives in the attention we give to the activity itself. Whether fishing or photographing, life is being lived as an art. Art is being made out of life.

The more we do something carefully, attentively, the more beautiful that something becomes. Writing about art in everyday life, the art critic Michael Kimmelman tells us to be alert to the senses. "Try to treat everyday life, or at least parts of it, as you would a work of art." Then you will experience the sublime, the spiritual awareness that comes with life intensely lived. Life, everyday, ordinary life, is our devotion.

Like a background noise in the universe, the war in and against Iraq continues to dominate the news—and easily our lives. News that cannot be ignored as we walk through the woods in the country and on the streets in town. An unnecessary war for corporate interests and government schemes of global domination. War atrocities, the killing of Iraqi citizens, death of servicemen and women, all these continue, and light is yet to be seen at the end of the tunnel.

As the American Civil War was about to begin, as slavery was being questioned in word and deed, Walt Whitman wrote a long poem titled "A Boston Ballad," a poem that would be part of his *Leaves of Grass*. Within the poem, he lamented "O ME! O Life!"—listing sorrows within the nation. And he wondered about his sorrow and what response might be appropriate. His answer was:

> That you are here—that life exists and identity,
> That the powerful play goes on, and you may contribute a verse.
> A verse, a line now and then, a moral response to the troubles of the times.

Whitman followed "O ME! O Life!" with the poem "I Sit and Look Out." For the record, here is the poem:

> I sit and look out upon all the sorrows of the world, and upon all oppression and shame,
> I hear secret convulsive sobs from young men at anguish with themselves, remorseful after deeds done,
> I see in low life the mother misused by her children, dying, neglected, gaunt, desperate,
> I see the wife misused by her husband, I see the treacherous seducer of young women,
> I mark the ranklings of jealousy and unrequited love attempted to be hid, I see these sights on the earth,
> I see the workings of battle, pestilence, tyranny, I see martyrs and prisoners,
> I observe a famine at sea, I observe the sailors casting lots who shall be kill'd to preserve the lives of the rest,
> I observe the slights and degradations cast by arrogant persons upon laborers, the poor, and upon negroes, and the like;

All these—all the meanness and agony without end I sitting look
out upon,
See, hear, and am silent.

Sometimes, more times than not, we sit—or stand—in silence. Silence being better than the actions and the noises that bring the war.

Here in the midst of the biggest snowfall of the winter, bound at home, one spontaneously turns attention to the wonders of this existence. No need for the ritualized formalities of traditional religious services. Our worship is in the awareness, in the mindfulness, that we give to the living of our daily lives. Life on earth as it might be in heaven.

If one draws from the New Testament for instruction, it is to the epistle of James (2:24): "A person is justified by works and not faith alone." This instead of Paul's words: "A person is justified by faith apart from works" (Romans 3:28). We are to help others, to love our neighbors as ourselves, knowing that the heirs to the Kingdom are the poor.

The prophet Amos (5:21–24) of the Hebrew Bible proclaims:

I loathe, I spurn your festivals,
I am not appeased by your solemn assemblies.
If you offer me burnt offerings—or your meal offerings—
I will not accept them;
I will pay no heed
To your gifts of fatlings.
Spare me the sounds of your hymns,
And let me not hear the music of your lutes.
But let justice well up like water,
Righteousness like an unfailing stream.

This Harold Bloom reminds me in his book *Jesus and Yahweh*, in a reading that has revealed much to me this winter season.

Deeds performed daily in place of a Sabbath of standing for hymns, for those who are able, and offerings to keep the edifice from falling down. Give us this day our daily life. Faith is in the lives we live with care.

As if this world were not enough. A new world is always promised in the monotheistic religious tradition that has prevailed these many centuries. The New Testament ends with the coming of that new world—a world that in no way is a natural development of this first world, but a new creation. Revelation (21:1–2): "Then I saw a new heaven and a new earth; for the first heaven and the first earth had passed away, and the sea was no more. And I saw the holy city, the new Jerusalem, coming down out of heaven from God, prepared as a bride adorned for her husband." Followed by a loud voice from the throne saying, "Death will be no more; mourning and crying and pain will be no more, for the first things have passed away."

This negative view of the world, and this hope for salvation in another life, gives me little comfort even in difficult and trying times. This view—this apocalyptic view—comes to us from a time long ago, over two thousand years ago, in a little place at the end of the Mediterranean. Now this is a literary genre, in scriptures Old and New, which gives definition to the lives of a good portion of the world's population. The book of Revelation, infused with earlier scriptures, written in a concrete time and place, provided solace to those threatened and persecuted by the Roman Empire in the first century. We today—especially in these current days—know the strong hold of the apocalyptic on global events, national debates and policies, and on our everyday lives.

Why make a better world here when one believes that we are on the way to heaven as this world ends? The eschatology being that this present and temporary world is certain to end, and that this world, dominated by the power of evil, will be gloriously replaced by the heaven of a God supreme. The final climax of human history will be, accordingly, the New Jerusalem, a world veiled by the tears of this earthly existence. Then wage the war in Iraq, killing citizens and troops alike, because the end is soon to come anyway. Why invest in programs for health, education, and welfare, taking resources from the rich, when the end is near? You would have to believe that heaven is some place beyond and outside of this earth to accept such reasoning and action.

As we transfer our vision of heaven to life here on earth, the apocalyptic ideology and spirituality will be relegated to an ancient and obsolete history. A kingdom here on earth comes in the struggle to make a better world here on earth. In the only world we can know and have our being and our hope.

What is the ground upon which we stand, the ground upon which we act? The elementary ontological question of the nature of reality remains a pressing one. Everything else follows from our understanding—our stance—toward the question of reality. What is real? Really real, actually existing. Simply to ask is to realize that reality is existential. All human perception is subject to the lived experience of everyday life.

A snowstorm followed closely behind the small jet that took me across Lake Michigan. I would be lecturing for the week at the university in Kalamazoo. I would be speaking about the circles that we humans, in our time and place, draw around ourselves and others, about the prisons we create whether we are inside or outside the institutions of incarceration. Our thoughts and discussions fast turned to things about knowing what is real in times of trouble.

Some years ago, during that ontological shift of the 1960s, I sat in the cafés of Greenwich Village, walked the streets of New York, and participated in the political events of the time. Daily, without ceasing. The song was playing, Les McCann's "Compared to What." Singing of the events of the time, hoping to understand, McCann sang, "Everybody now, try to make it real, compared to what?" He sang mournfully the lines: "The goddamn nation ..." "The president he's got his war. Folks don't know what it's for." "But I can't use it." "Where's my God, and where's my money?" "Goddamn it, try to make it real compared to what?"

In those days, I was trying to make my own song. One result was a book I called *The Social Reality of Crime*. Shortly after its publication, Taylor, Walton, and Young, in their influential book *The New Criminology*, wrote about my efforts: "Many of Quinney's statements about a theoretical orientation to the social reality of crime seem to be the product more of the author's own existential angst than they are the result of clear-headed theoretical analysis." I remain to this day pleased with their observation. For one thing, I am happy to be counted among the existentialists. Albert Camus, I think of you daily. ("Mother died today, or maybe it was yesterday.") And, "clear-headed theoretical analysis," abstract and removed from everyday life, is not something to which I aspire.

My questioning of the conventional scientific enterprise took further attack from Robert Merton in his book *Sociological Ambivalence*. He wrote:

> Now, it is one thing to maintain, with Weber, Thomas, and the other giants of sociology, that to understand human action requires us to attend systematically to its subjective component: what people perceive, feel, believe, and want. But it is quite another thing to exaggerate this sound idea by maintaining that action is nothing but subjective. That extravagance leads to sociological Berkeleyanism (the allusion being, of course, to the English champion of philosophical idealism, not to an American geographic or academic place). Such total subjectivism conceives of social reality as consisting only in social definitions, perceptions, labels, beliefs, assumptions, or ideas, as expressed, for example, in full generality by the criminological theorist, Richard Quinney, when he writes that "We have no reason to believe in the objective existence of anything." A basic idea is distorted into error and a great injustice is visited upon W. I. Thomas whenever his theorem is thus exaggerated.

I still do not know if I am a total subjectivist, or even what that might be. But it does seem to me that the subjective/objective debate is as much ideological as it is a statement about what is real and how we can know reality. The problem goes beyond a debate over the subjective and the objective. The matter has something to do with the human mind's ability to think and to see beyond received conventions.

How can we know for certain the existence of anything, including existence itself? The mind is the grand piano, which provides the space for the mice—our thoughts—to play. We humans cannot step outside of our existence. And we cannot know, in the larger scheme of things, or non-things, if the grand piano is other than a dream. The dream of a cosmic dreamer.

It is not for us to know that which cannot be known. To seek such knowledge is not to be human. The simple teaching of Buddhism: "Only don't know." We have the mind to ask questions of the reality of our existence, universal and otherwise, but we do not have the capacity to answer with objectivity and certainty. (Camus, again: "The absurd is the essential concept and the first truth.") Entirely reasonable, then, our perpetual ambivalence, our uncertainty, and our fear of life and death. Humility, mixed with wonder, makes more sense than the obsessive search for scientific knowledge.

We stand before the mystery of existence. Our understanding is in the recognition of our common inability to know for certain. Our fate, and our saving grace, is to be compassionate beings, in all humility. Whatever may be known is known in love. Not in manipulation and control, not in the advancement of a separate self, but in care for one another. That is reality enough.

If reality is enough, why is there the desire to write—and to read—fiction? Perhaps fiction is another way of trying to grasp reality. Of course, all narratives are, from beginning to end, fiction. As soon as a writer begins to describe even what is going on in the concrete moment, a new reality is being created. The writer in the elementary act of selecting a subject is imposing an order that is at once a remove from the event that is being recorded. And the narrator is always writing from a unique perspective that has been developing over a lifetime. All is fiction, which makes the writing of "nonfiction" all the more a moral act and a labor of great responsibility. We construct the realities by which we live our lives.

In the preface to his book of essays *Vermeer in Bosnia,* Lawrence Weschler makes this pertinent observation about nonfiction writing:

> The part of my sensibility which I demonstrate in nonfiction makes fiction an impossible mode for me. That's because for me the world is already filled to bursting with interconnections, interrelationships, consequences, and consequences of consequences. The world as it is is overdetermined: the web of all those interrelationships is dense to the point of saturation.

Thinking that you are writing about what is real, about the complexity of reality, leaves little room or need for fiction. There is no need to try — to pretend — to be somewhere else.

I have scant need, for whatever reason, to invent a world beyond the one I am experiencing. Edward Hoagland has this to say about the writing of nonfiction, as opposed to writing a novel:

> Novelists want the site of their drama to be ground zero, but most of us do not live at ground zero. Most of us live like stand-up comedians on a vaudeville stage — the way an essayist does — by our humble wits, messing up, swallowing an aspirin, knowing Hollywood won't call, thinking nobody we love will die today, just another day of sunshine and rain.

My "nonfiction" is a recognition of my own mortality and my own precarious and humble hold on the life I am given.

At the end of a brief essay on the life of nature writer Edward Abbey, Hoagland quotes a passage from Abbey on living close to the earth: "Some of us have the nerve, the insolence, the brass, the gall to whine about the limitations of our earth-bound fate and yearn for some more perfect world beyond the sky. We are none of us good enough for the world we have." It is enough to live in and to appreciate this world here on earth.

On a dark night in the country, away from the bright lights of the city, superstitions could easily surface. I imagine that down at the old place, as John and Bridget watched the last embers of the fire fade, the little people of the whitethorn forts in County Kilkenny could be remembered. Without the artificial light that would come later, night fell early in the wintertime. Drifting in and out of sleep all night long, dreams and reveries brought back the fairy people. Even now as I pass the old place with its foundation of crumbling stone, I sense the mystery of another time and another place.

Much could be explained, before the advent of electric light and the radio, by the presence of fairies and by superstitions of various kinds. Celtic customs survive in the deep and subtle recesses of my mind and spirit. Especially when I return for a night of restless sleep in the farmhouse. I will let the fairies do their work and give my attention to other matters in the light of day.

Once, some years ago, I traveled to Ireland and stayed several months to explore and to be near the source of my Irish heritage. With my family, I drove and walked the narrow roads, tried to heat the house with turf, and went to dances in the local pub at night. I taught classes at the university in Galway, teaching the rebellious theme of Stop Making Sense.

Upon our arrival in the rented cottage in Annaghdown, on a winter's night, my daughter read my future with Tarot cards at the kitchen table in the dim light of the kerosene lamp. I asked,

"What is the nature of my work?" Selecting the Magic Seven Spread, Anne lays out the cards: Pope, Moon, Three of Coins, Magician, Eight of Cudgels, Tower, and King of Coins. I learn that I am indeed in a dark time. But in the near future, I will come into a new phase that will be enhanced by being in a magical place. There will be contentment. The last card, the King of Coins, has the character of Gemini: "You should budget your energies so that a scattering of forces will not dissipate your true talents." Gemini, I am told, is not bound by material motives. The cards advise me: "Instead of changing occupations, you should use your mental agility to discover new methods in the one you have chosen." The environment of Gemini is where the wind blows and the land is rolling and covered with grass.

While in Ireland, I read Sean O'Faolain's book *The Irish*. I read that a sense of the Otherworld dominates the Celtic imagination, with this imagination seeking "a synthesis between dream and reality, aspiration and experience, a shrewd knowledge of the world and a strange reluctance to cope with it, and tending always to find the balance not in intellectual synthesis but in the rhythm of perpetual emotional oscillation." The next day in class, I read aloud a passage from W. B. Yeats, from an introduction to the fairy stories he gathered in the west of Ireland: "I have always sought to bring my mind close to the mind of Indian and Japanese poets, old women in Connacht, mediums in Soho, lay brothers whom I imagine dreaming in some medieval monastery the dreams of their village, learned authors who refer all to antiquity." Yeats adds, referring to expectations for his own mind, "To liberate it from all that comes of councils and committees, from the world as it is seen from universities or from populous towns." I tell the class of my own attempts to go beyond the modern discursive mind, to open the mind to the mysteries that are unknown to conventional thought.

Down the road from our cottage, pieces of broken mirror glass border the front doorway and windows of a farmhouse. A neighbor woman tells me this mirrored glass will keep the "evil eye," Old Nick the devil, from entering the house. In the evening, John Dooley, a farmer farther down the road, comes to our house for a visit. I ask him about the marsh that runs along the farmland. He asks, "Have you heard about the fairies?" He then tells us the story about the farmer near the marsh who had brought a log from the marsh and had placed it beside the fireplace where it remained for twenty-one years. One night, a galloping horse came by the house, and the fairy rider shouted, "Hurrah, Hurrah. Did you see Dermot?" The log rolled out of the house and onto the horse behind the rider, and they galloped back to the marsh. John Dooley told us that he does not himself believe in fairies, but that there are others nearby who still do. The next day I walked to the place where the incident had occurred, and I viewed the remaining foundation of the abandoned farmhouse.

In the creation myths of some tribes, it is turtles all the way. One turtle on the back of another all the way back to the origin of things. Something like this comes to mind each time I stop at the bottom of the hill to let a turtle pass over the road. A turtle following the ancient path that goes from the muskrat pond to the marsh on the other side of the road. With the passing of years, I have seen fewer turtles crossing the road. There is some hope that the turtle population will increase as the wetlands are restored.

I now think of Gilbert White, the English curate living in

the village of Selborne in the eighteenth century, of his keeping a turtle—actually a tortoise brought to England from the Mediterranean—and making his observations of the tortoise throughout the seasons of many years. White stayed close to home to make his observations and write his letters, to pursue his studies of "the life and conversation of animals."

Timothy is the name White gave the tortoise. (Although in the end, after the death of Timothy and transfer of the shell to a museum in London, it was determined that Timothy was a female, not a "he.") White wondered, most of all, why the tortoise should be granted such a long life. In a letter to a friend, White exclaimed: "Such a profusion of days, such a seeming waste of longevity, on a reptile that appears to relish it so little as to squander more than two-thirds of its existence in joyless stupor, and be lost to all sensation for months together." Nonetheless, or maybe because of such wonder, Timothy was named and cared for as an individual rather than being merely of another species.

What a surprise and delight to read the newly published book *Timothy; or, Notes of an Abject Reptile* by the nature writer Verlyn Klinkenborg. It is a story told from the point of view of Timothy, Gilbert White's beloved tortoise. This time Timothy is the observer and narrator, and White and his family and friends are the ones being observed and written about. Timothy ponders the big questions of existence, the ones that even escape the curate. Timothy knows that he is more than a "walking tureen of turtle soup," more than "a meal gone missing." And of White, Klinkenborg notes: "The real question he knows. Why any creature is allotted time on this earth. No matter how brief or how long. He knows this. Even if he hides the question from himself."

Timothy tells us of his own sensibilities that are beyond human apprehension:

> I attend to the shadows. To the night-shapes of the trees surrounding this garden. Thatched bulk of houses down the village-back. I cannot hear everything the foxes hear or smell everything the dogs smell or fear everything the humans fear. But I can sense the remote static of the aurora borealis stretching east to west across the welkin. Feel the stars overhead when the wind drops. Resonant in their silence. Smell the dawn when it is still just a premonition. A stirring in the throat of the earliest songbird. The wakeful rooster.

In the end, Timothy wishes to be out of human reach, to leave this human-bound world.

> But my motive for leaving is plain enough. Neither hunger nor love. I wish to be out of human reach. Out from under the constant stir. Laborious turmoil of this breed. Endless bother of humans. Toil inherent in their mere existence. Dizzying inability to bask or muse.

Timothy, the tortoise, will pass to a world beyond human reach and human perception. Beyond the limited mind of the human being. "I wish to live again in a place that is not a map of the gardener's mind. Book of nature, as humans love to think of it. But where I wish to live is not a book at all. Not an argument for the being and attributes of an unnecessary god. Not a theorem, hypothesis, or demonstration. Merely itself." Might that we see, however dimly, an existence beyond that known to our human being.

If there is a force in the universe, one that we dare call "God," and one that is benevolent toward our own species, why do bad things happen to us? Most of all, why must we die? A force, an ultimate being perhaps, that favors us would not let us die. If there is an order in the universe, even one based on chaos, it does not favor humans over all other things. Something is going on that is beyond our human interest and human understanding.

The scientific journals and the newspapers are reporting that within the next one hundred years sea levels will rise by twenty feet. The rise in sea level is likely caused by human influence on Earth's climate. Vast polar ice sheets are melting, average temperatures are rising, and ocean currents are changing, in part at least, because of emissions of carbon dioxide and other gases that are trapping heat in a greenhouse effect. The recent flooding of New Orleans is probably a signal of things to come as a result of global warming. At what point, the question is being asked, will this process become irreversible if something is not done to curb the warming of planet Earth?

Sea levels have been rising naturally for thousands of years following the Ice Age that ended ten thousand years ago. Certainly we humans have accelerated the warming of the planet. The seas are rising, and we are among the many that will be seeking a higher ground.

Warming of this planet Earth is but a small event in cosmic terms. Across the universe, as we speak, there is Crash, Bang, and Boom. Planets collide and meteors and asteroids crash into stars and planets. Dinosaurs and most other species once disappeared from the planet, becoming extinct, because of one such impact sixty-five million years ago. An asteroid seven miles in diameter ploughed into the planet near what is now the

Yucatan Peninsula, baking the planet at five hundred degrees, and blocking out the sun as blackened soot rose from the ground. There were a few survivors, enough to begin a process of evolution that would include our ancestors.

This could happen again. But there is a much bigger collision that will eventually happen. A collision that is out of our human ability to avoid. Billions of years from now, the Milky Way galaxy, where Earth and Sun reside, will collide with the Andromeda galaxy. These two galaxies, some two million light years away from each other, are moving toward each other and will eventually collide and merge into something new. Gases will compress and heat, and countless stars and planets will be created. New arenas for life will form, just as a collision preceded us and made our lives possible. And then someday, that universe will sail on to oblivion.

One does not need to posit a divine apocalypse. In the larger scheme of things, there is a natural end to the world. But we humans cannot live daily thinking about the apocalypse, divine or natural. Let any awareness of the larger scheme of things make us humble and caring in our daily lives. There is enough violence in the natural world. Let our human world be one of peace.

Each life is a universe unto itself. A big bang. Sperm collides with egg. There you are. Then a lifetime of galaxies and stars and planets and asteroids flying and orbiting in a vast and expanding space. More dark matter than light. Shiva dancing. A great mystery. And a beautiful spring morning it is this day.

I keep notes along the way. Now and then an accounting and some reflecting. No remorse, more wonder than remorse, and gratitude for the journey that continues. Thankful for all the parallel universes that have come my way.

I am struck this day by the words of Edward Hoagland at the end of a chapter in his memoir *Compass Points*. Noting his hopes and dreams and the realities of his life as he ages, he writes the following:

> I wanted America to be classless, as I thought a democracy should be, but since no utopia was in the offing, I put my energy into books, not changing things. As in a marathon, everyone hits a wall at some point, and you want to reach where you hoped to wind up at that time of life: to "have written your books," as writers say, and done what you could, when the stroke hits or you simply run dry.

A sense of reaching the finishing line. You know that one day, probably without warning, the line will be reached. The universe, your universe, will end. Although it will no longer matter to you, and you will no longer fear death, it matters to you now that you "have written your books," that you are writing your books before running dry or before the stroke hits you. In other words, that you are alive until your universe comes to its natural end.

Always the return to place, to a particular place. A return, most often, to the place of origin. The latest reminder is an essay in the current issue of the Buddhist journal *Inquiring Mind*. The author, Susan Murphy, a Zen teacher in Australia, writes:

> We cannot but live somewhere in particular, a particularity that our bodies register and express so intimately that it is finally impossible to say that it is not our own body. The air we breathe

and feel on our skin, the water we drink and breathe in, the smell of the soil and the plants springing from it, the weight of our footsteps and the answering pressure of the Earth, the shape of the hills, the light of the sky. To overlook all this and to live in vagueness toward where we live, the place that lives us, is to place our living place in peril.

We constantly renew our awareness of our sense of place. And this renewal—this affirmation—is a part of our practice, spiritually and materially. Our meditations are ritual acts of respect for the place—or places—in which we live. The Zen teacher reminds us, "Reality is connectedness. If you're not connected, you're not *in* reality."

And when we travel from our place, this travel is also part of our practice. A friend from a far northern corner of Minnesota has written to me of such travel. She writes: "But what is travel? In its most elemental form, I think travel must be rest. I think we must rest our minds of place in order to know place. We must rest our minds of our immediate experience in order to acquire a fuller perspective of our experience." What may be the nature of this travel? "A walk to the marsh, perhaps; a geographical shifting from one place to another; an afternoon's meditations; a change of company; a transcending of the immediate in order to attain a sense of the whole." In closing, I am given a gentle wish for a fine day.

Lush is the earth, fecund and fertile. Lilacs spread profusely across yards in the Midwest. Tall lilac bushes surround the crumbling foundations of houses on abandoned farms. And in our field at the farm, the new foal suckles the Fjord mare Gunnel. Frey is his name, after the son of the Norwegian god of fertility. On his feet just one hour after being born in the field. Very near the date and place of my own birth many years ago.

In this world, of course, for every birth there is death. The morning paper carries the obituary of the poet Stanley Kunitz, dead at the age of 100. The obituary notes that he shunned shallow confession in his art. Kunitz had written: "Poetry is ultimately mythology, the telling of stories of the soul. The old myths, the old gods, the old heroes have never died. They are only sleeping at the bottoms of our minds, waiting for our call." In an interview in *People* magazine a short time ago, he said, "The deepest thing I know is that I am living and dying at once, and my conviction is to report that self-dialogue."

I am living the years that my father did not live. Now I am three years older than my father was when he died. Kunitz wrote in a poem:

> Peace! Peace!
> To be rocked by the Infinite!
> As if it didn't matter
> which way was home;
> as if he didn't know
> he loved the earth so much
> he wanted to stay forever.

Someday it will not matter that I too wanted to stay forever.

The message has been hidden for sixteen hundred years. The Gnostic gospel was found in an Egyptian cavern only a few years ago. Now, after scholarly reconstruction and translation from the Coptic, we can read and ponder its significance.

Judas, the disciple of Jesus, has been known for centuries as the betrayer. In the newly discovered gospel, Judas is the one who made possible the transcendence of Jesus from this material existence. And in the gospel, as one of the gospels excluded from the Bible as heresy, there is another world apart from the one offered to mankind by the God of the Old Testament. In the Gospel of Judas, in the Gnostic tradition of the secret knowledge, there is a world beyond the one of the lesser god portrayed in the Judeo-Christian tradition. Passage after passage from the interpretive essays in *The Lost Gospel of Judas* take the reader to places opposed to and beyond conventional knowledge.

The Gospel of Judas is a narration of an entire cosmological myth, ending with the creation of humanity by inferior gods. According to this gospel, written originally in the second century, and true to Sethian gnosticism, Jesus is not the son of the God of this world, but is from the realm of Barbelo. Judas declares to Jesus: "I know who you are and where you have come from. You are from the immortal realm of Barbelo. And I am not worthy to utter the name of the one who has sent you."

In the immortal realm, called Barbelo, is the true origin of light and life. Judas is the disciple that aids the son (Jesus) of Barbelo to rid himself of the material body and enter the world beyond the one created by lesser gods.

Such is the good news of the Gospel of Judas. The news is of an immortal realm beyond the disasters of this world of earthquakes, floods, famines, epidemics, and suffering. This mortal world here on earth is not a home for anyone with the knowledge and spark of the soul coming from the immortal realm.

Whether we humans are predestined to enter that realm (or not enter) after death, may depend on the form of gnosticism that is being entertained. Outside of the problem of predestination, or even in spite of it, the gnostic perspective of Judas is likely to expand our consciousness. Why we are here and where we are going can never be known by this human mind. To this extent—this big extent—we are clueless.

I will tell you a story. Twenty-some summers ago I traveled to China. A journey to the other side of the world seemed important for reasons only vaguely sensed at the time. Near the end of the journey, I asked a fellow traveler whether it is necessary for us to travel. I had quoted to her lines from the *Tao Te Ching:*

> Without going out of my door
> I know all things on earth.
> Without looking out of my window
> I can know the ways of heaven.
> For the further one travels
> The less one knows.
> The sage therefore
> Arrives without traveling.
> Sees all without looking
> Does all without doing.

In a letter to me after returning home, she answered: "It seems we are always traveling as a part of living, although we become most conscious of this only when we arrive or depart, when we are attuned to or watchful of things newly seen." She added, "I will miss you as a traveling companion."

Preparations for the trip to China were simple and few. I read the briefing materials supplied to the twenty-five delegates who would investigate law and justice in several cities and provinces of China. I learned to play a piano rendition of Frank Loesser's "On a Slow Boat to China," and watched Willie Nelson's video about the songwriter: "Write it down—what you found—songwriter. Don't let it slip away. Someone is listening today." I read a book of love poems by Chinese women of the early dynasties. I sat with Cambodian monks at the Buddhist Temple and listened to the Dharma talks of Maha Ghosananda.

When asked by friends why I was making the trip to China, I quoted the words of Chuang Tzu, written over two thousand years ago. "The good traveler doesn't know where he's going. / The great traveler doesn't know where he's been." Pressed further, I replied simply that I was going to China to see what it is like to go to China. To observe the elusive legal system in three weeks would be like, according to the Chinese aphorism, "looking at flowers from a galloping horse." The trip would be a continuation of my own journey of bringing Eastern sensibilities and practices into my everyday life.

High over the continent, jet engines roaring, the airplane took us on the first of many flights. "In the place where we are now," Alan Watts once said, "we have already arrived." I write a postcard to a friend, quoting the Third Chinese Patriarch of Zen, "The Great Way is not difficult for those who have no preferences." The airplane descends through the clouds and rolls smoothly down the runway to the San Francisco terminal.

We spend the night in a hotel in San Francisco's Chinatown. In the drawer of the bedside table is a copy of *The Teaching of Buddha,* placed next to the Gideon Bible. I am told by housekeeping that I may take the copy with me to China. Tai Chi is being practiced on the walkway outside the hotel window as I get up the next morning. We repack our suitcases, meet the bus at the lobby door, and return to the airport where we board the large 747 that will take us to Tokyo and, after a brief stop, on to Beijing.

Trees line the narrow road from the Beijing airport into the city. In the middle of the night, we pass slow-moving bicycles in the yellow misty light of street lamps as our bus makes its way to the guest hotel. Horse-drawn wagons for the collection of night soil are making their rounds. A Buddhist book had reminded me "we are all of us interrelated—not just people, but animals too, and stones, clouds, trees."

In the morning we make our visit to the U.S. Embassy. Between the balsam poplars of Ritan Park, cicadas singing into the moist air, we walk to the compound for our meeting. Seated in the large room, with a fan slowly moving overhead, our delegation is eager to begin asking questions. Embassy officials have their own concerns: apprehension about economic reforms, anxiety over relations between China and Russia, and fear of another cultural revolution. Already I am attuned to another wisdom: "Within Siddhartha there slowly grew and ripened the knowledge of what wisdom really was and the goal of his long seeking. It was nothing but a preparation of the soul, a capacity, a secret art of thinking, feeling, and breathing thoughts of unity at every moment." At this moment, I can think of no questions that need to be asked.

In the afternoon, we begin our meetings with officials. We are seated around a large table and in stuffed chairs at the edges of the room. Teacups have been set before us, and soon young women come carrying enormous thermos bottles and pour hot tea into our cups. The leading official begins his prepared statement. These meetings will be repeated in similar settings and fashion throughout our trip.

There seems to be a wide gap between the questions and answers. English to Chinese and back again to English, problems of translation are obvious. My listening turns to the sounds of the language and the songs of the cicadas that fill the room. The truth is somewhere beyond this ceaseless search for knowledge. Later in the evening, a few of us walk through Tiananmen Square and watch the children and families in quiet play. Nighthawks fall from the dark sky.

As bicycles fill the morning street, I reach for the *Teaching of Buddha* and turn to the section on "Human Defilements." The passage reads: "There are two kinds of worldly passion that defile and cover the purity of Buddha-nature. The first is the passion for analysis and discussion by which people become confused in judgment. The second is the passion for emotional experience by which people's values become confused." Our bus stops at the Memorial Hall and we file past the body of Chairman Mao. Outside, workers in straw hats dig in piles of sand, and cranes are extended high above scaffolded office towers. We move on, asking more questions and seeking new experiences.

I give my copy of *The Dhammapada* to a friend as we approach the Temple of Heaven, and we read: "Find joy in watchfulness; guard well your mind. Uplift yourself from the lower self, even as an elephant draws himself out of a muddy swamp." Meditatively, without choice, with what Buddhists call "choiceless awareness." We quietly enter the circular Hall of Prayer for Good Harvests. In watchfulness we behold the treasure of awareness.

At the Beijing airport, on our way to Inner Mongolia, a young schoolgirl approaches as we sit waiting for our flight to be called. She asks me, "From where do you come?" I answer, and then she tells me that she is in the lobby practicing English with foreign visitors. Her hope is to become a translator for an agency of the Chinese government. After a delay caused by the hot night air, we board the propeller airplane to Hohhot. Each of us is handed a paper fan for the flight. As we rise into the sky, condensation from the air system appears like smoke in the cabin. We fan rapidly into the night sky as the plane flies north over the Great Wall.

In the morning, we are on the bus for an excursion over the Daqing Mountains to the grasslands of Inner Mongolia. Our tourist guide, dressed as a Mongolian herdsman, comes galloping over the horizon of China's frontier, and sings a native song as he dismounts. We are shown a ritual site on the top of a low hill, the plains spreading out in all directions with a few adobe farmsteads dotting the landscape. In a village, we walk into the courtyard of a beautiful, brightly colored lama summer temple of the Living Buddha. Inside the temple, an aging lama appears out of a dark corner, and at the altar chants an ancient scripture. We place coins at the altar, and I remain for a brief meditation. Later in the evening, we dine on mutton cooked in hot pots, and are toasted lavishly by our Inner Mongolian hosts. We gather in a yurt for a night of sleep under the star-filled sky.

Two days later, after returning to Beijing on a train pulled by a steam locomotive, we fly the two thousand miles to the far western province of Xinjiang. Arriving in Urumqi, we sort luggage and are given rooms for the night. This is a time to practice peaceful existence within the confusion of daily life, practiced in a far corner of China. But soon we are traveling in the heat of the day over dusty desert roads to the Heavenly Lake in the Tian Shan Mountains. Later in the afternoon we stop in a farming village, and gather around tables placed beside a village house.

Overhead ripening grapes hang from an arbor; tea is poured, and watermelon slices are served. Children in bright dress clothes watch our proceedings from a distance. An occasional swallow darts in and out of the windows of the house. Incense burns from under the eaves.

The village is composed of a thousand households of different ethnic groups. The village chair welcomes us, and soon our delegation is asking questions that are becoming repetitious. My lasting impression is of the sounds and smells and sights on a late afternoon in a Chinese village in a far province. The villagers take us on a walk around the village before the dinner that they have prepared for us. I write a lasting impression in my notebook:

> Evening under the grape arbor,
> Seeking to know the ways
> Of Han, Hui, Uygur, and Kazakh—
> Children watch from a distance.

The dry and dusty air of Urumqi produces an allergic reaction that requires me to remain in the hotel for the rest of our stay in Xinjiang province. The guest hotel has the look of a monastery, and I welcome the opportunity to be alone in a spacious room while the rest of the delegation goes on a two-day sightseeing trip into the desert. I wash clothes and walk slowly down the wide hallways over the polished yellow and brown floor tiles. Well-worn oriental carpets cover the lobby of the aging hotel. I read in the stillness of my room the words from *The Dhammapada:* "Whereas if a man speaks but a few holy words and yet he lives the life of those words, free from passion and hate and illusion—with vision and a mind freed, craving nothing both now and hereafter—the life of this man is a life of holiness."

A gentle rain is falling as we arrive in Shanghai and continues during our days in the city. At night the brightly lit boats along the Huangpu River reflect on the water and into the mist as we walk the Bund. Couples openly show their affection. We are told in meetings that the streets of Shanghai are now free of crime and prostitution, and that the old era of slums, opium dens, and starvation virtually ended with the reforms brought about in 1949 by the People's Republic of China.

All over the city, billboards advertise consumer goods: television sets, motorbikes, kitchen appliances, and cameras. "Red-eye disease" afflicts those who desire more than they can afford, we are told by an official. He tells us, "Outside influences cause people to make themselves satisfied without regard for others." Returning to the Peace Hotel from a banquet of many dishes, we listen while our guide from Shanghai sings for us "We Shall Not Be Moved."

As we land on the outskirts of Fuzhou, the coastal capital of Fujian province, we see water buffaloes pulling wooden plows through rice fields. High on Drum Hill, Buddhist pavilions and temples are being repaired and rebuilt. We visit an auto-works factory commune and witness the mediation of a dispute between a mother-in-law and her daughter-in-law. Walking into a hotel bookshop, I find a volume of the poetry of Wang Wei. I copy into my notebook the line "I live as I please, unshackled by external fetters." Even in this faraway land, I seek some remove from worldly matters.

In Guangzhou, we are housed in the luxurious White Swan Hotel. Across the Pearl River is Shamian Island, once the British enclave made famous during the Opium Wars. Once signs in the compound of large Victorian buildings warned "No Chinese and Dogs Allowed." We are being shown that a developing China can provide modern luxuries. During a Sunday morning session with the prosecutor of economic crimes, I look out the window and write these brief lines:

> Bridge over the River Pearl—
> Among the ruins
> The great white bird nests.
> High upon her back
> We watch ancient boats
> Speed toward the other shore.

Soon we will take the three-hour train ride to Hong Kong.

Over the Pacific, flying backward in time from Hong Kong to San Francisco, we watch as James Stewart pursues the elusive Kim Novak in *Vertigo*. The man next to me, as we near San Francisco and awaken from our sleep, tells me that he is a Chinese Buddhist now studying engineering in Philadelphia. In his own life, he tells me, he is trying to integrate what he calls the "passive" aspects of Buddhism with the "active" life of the West. But he cautions, "Always be attentive to the suffering that comes with the desires of human existence." Suffering can be stopped by the right practices, he tells me, as we depart at customs.

A flight to Denver, then on to Chicago, and finally a Greyhound bus ride to DeKalb, and I am nearly home. The sun is shining brightly and a morning breeze is stirring a sweet, grassy scent into the air. On the bus, I read the lines of Chang Tzu, advising me on how far I might want to go in understanding and explaining my trip to China.

If you persist in reasoning
About what cannot be understood,
You will be destroyed
By the very thing you seek.

I will continue to live in the mystery that is always present in this world of everyday life. The truth is already here, in this very moment, in this very place. An orange and black monarch butterfly flutters across my path and flies into the lilac bushes. This life, although in many ways simple, is a mystery that I will continue to welcome with each day's journey.

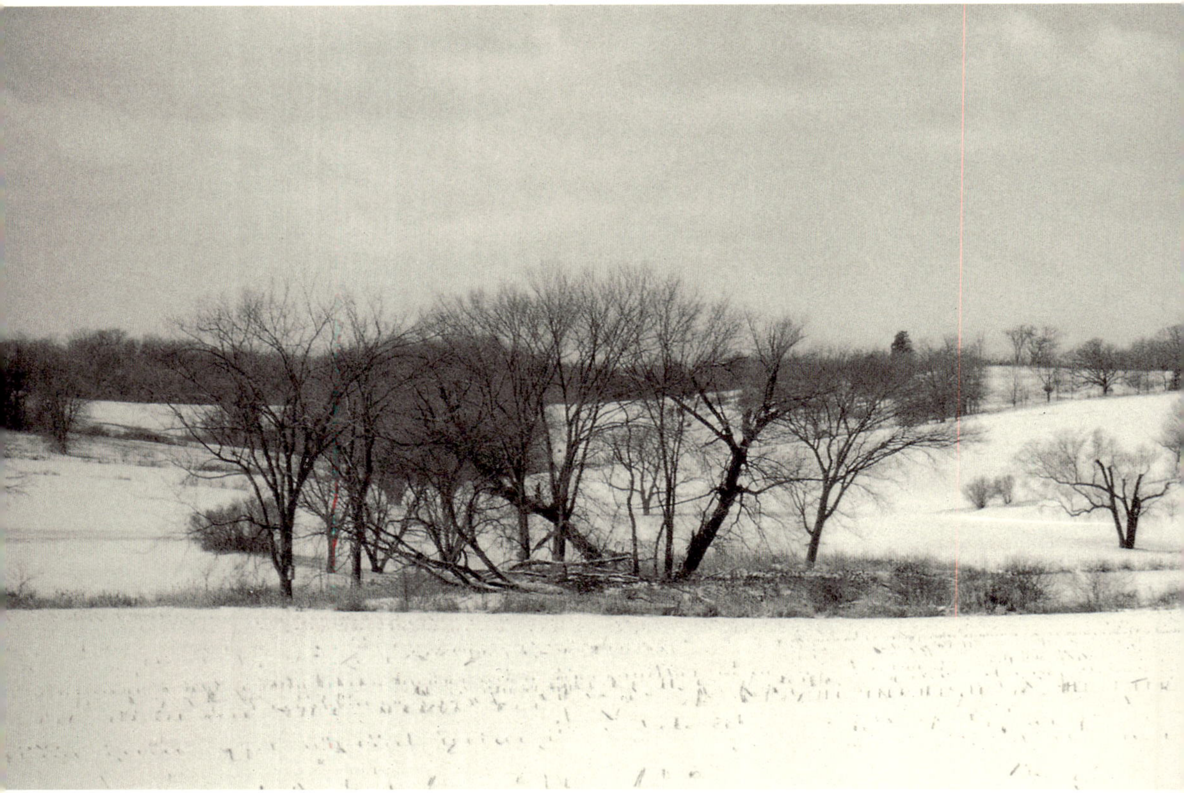

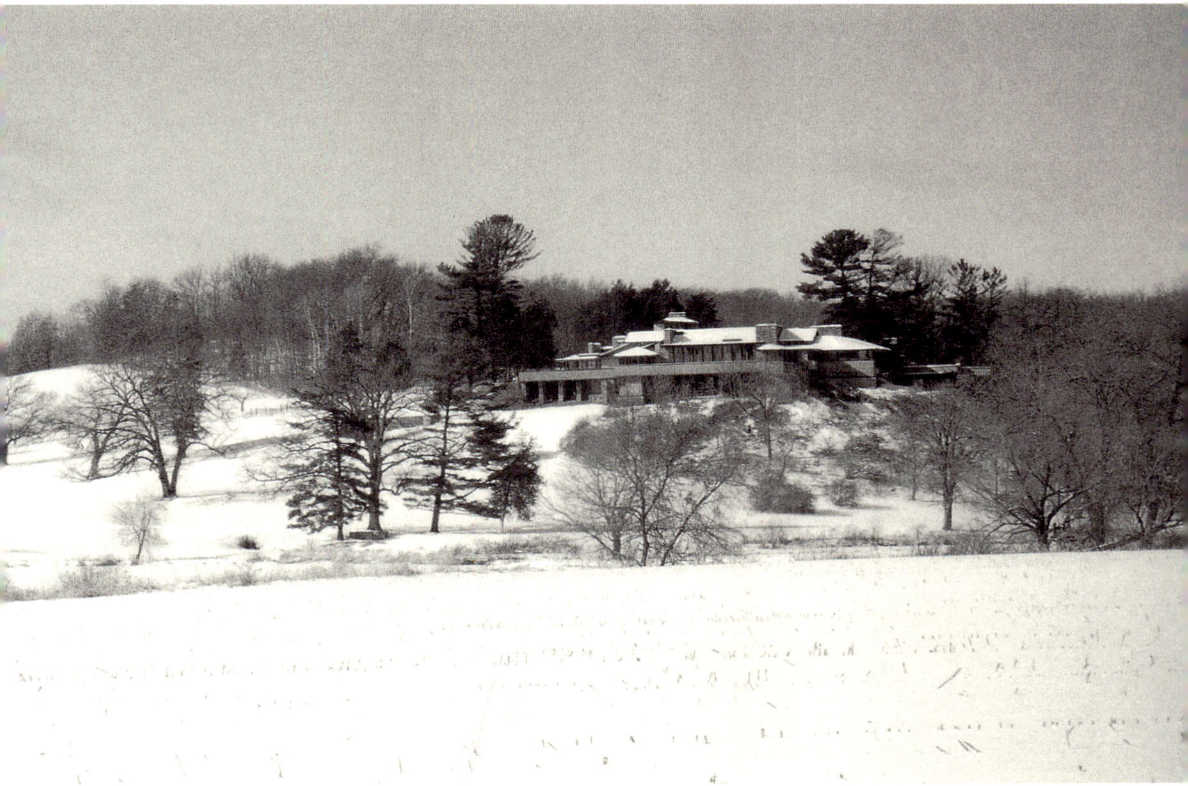

Part III

This Floating World

The moon is a house
In which the mind is master.
Look very closely:
Only impermanence lasts.
This floating world, too, will pass.
Ikkyu Sojun

The holy texts say that this world is an illusion. In a realm much higher than human consciousness, in a world of absolute reality, this may be true. But we live our everyday lives believing in the reality of our earthly existence. In our heads are the constant narratives—stories—of the lives we are living. Sketches from nature, the only nature we can know.

I continue to go to the front porch of the farmhouse to open the heavy lid of the emigration trunk, and dig a little deeper into the accumulated layers of artifacts from the generations of my family. As an archaeologist, of sorts, I excavate to learn more about the lives of those now gone. Gone from this world, illusionary or not.

Just this week, I found in the trunk the ledger kept by my

Grandfather Holloway for the years 1913 to 1921. On the pages are the finely written figures recording expenses and receipts. Glued to the inside of the front and back covers are clippings from local newspapers and farm journals. There are the practical articles on animal husbandry, one written by my grandfather, W. V. B. Holloway:

> I usually wean the little pigs at eight weeks old and let them suck their mothers every other day for about a week. This will often eliminate caked and swollen udders, which is so often the cause of tumors on the sow's udders. As soon as the little pigs will eat let them run to a self-feeder with ground oats, corn and barley or succotash as feed.

Among the practical matters of farming are various moral tales. Glued to the inside of the cover is a clipping inspired by Benjamin Franklin, titled "Let There Be Light." The article quotes the words Franklin offered to the makers of the Constitution: "The longer I live, and the more I know, the more I believe that God governs in the affairs of men, and, if the Sparrow cannot fall without His notice, is it probable that an empire can Rise without His assistance?" Next to this clipping is a chart on the "Quantity of Seed to Plant an Acre," followed by printed charts on the "Carrying Capacity of a Freight Car" and "Weight per Bushel of Grain," and short articles on "Worms in Horses" and "Feeding and Managing the Hog Herd."

Glued to the inside of the front cover of the ledger is a long printed column titled "Why People Save. The Ladder that Leads to Independence," by B. C. Forbes. The article begins with a question, and provides an answer:

> Why do people save money? Happiness, satisfaction, is what we all seek to get out of life. Is that not so? If you haven't thought about it, think it over now. Analyze yourself, analyze your feelings, analyze your motives, and you will see that you are actuated by this basic desire. The person who saves money calculates that he (or she) will get more satisfaction, more happiness in the long run through possessing the money saved than would be derived from spending it as it is earned.

Thrift is the ladder "which leads up to independence, to self-respect, to manly manhood." The author concludes, "And having thus calmly and deliberately made up your own mind, let nobody, let nothing tempt you into a side path leading to the almshouse."

A treasure trove of moral and practical advice from the past. On the last page of the ledger, my grandfather records the gain in height of my mother between April 28, 1916, and January 5, 1917. At eleven years of age, the girl who would become my mother grew from 52 1/2 inches in height to 55 inches. My grandfather also notes, on September 12, 1921, the death of his wife, Lorena: "Ma died." On the same page is taped a leaf, crumbling beyond identification. I have yet to thumb through the pages between the front and back covers of this tattered, cloth-covered ledger, a record of a time lived, and lives now gone. Another time I will dare to dig further into the trunk of this treasured past.

Living in two worlds—or between two worlds—has been the theme that runs through a good portion of literature, and through my life as well. "Two Worlds" is the theme that dominates the writings of V. S. Naipaul, and it is the title of the lecture he gave upon receiving the Nobel Prize for Literature. Whether you are a refugee, a migrant to another land, or someone who has moved from one culture to another, you are in two worlds. Part of your life is borrowed or defined for you by others. Naipaul, in an interview following the publication of his novel *Half a Life,* said that he escaped some of his own half-life by being a writer. A writer, he noted, invents a life in the process of imagination and in the construction of a narrative.

We are all between worlds of one kind or another. One of these divides, of course, is between science and the arts. Although much of my understanding comes from science, and the kinds of actions that follow such knowledge, it is in the realm of art and books and music that I gain another kind of understanding. It is out of this latter kind of knowing that I try to understand that other division of worlds: the world of the prison and the world outside the prison.

A society that separates people through the institution of the prison creates populations of incomplete and wounded lives, whether we are *inside* the prison or *outside* the prison. This is the dance of slave and slaveholder, inmate and captor, prisoner and non-prisoner. No one escapes the damage caused by the fact that the prison exists. The damage is pervasive—on levels economic, social, and psychological, and ultimately spiritual.

Although there are real and consequential differences between the lives of those in prison and those outside of prison, the injuries caused by the prison are shared by all. No one, in a sense, is an outsider. We are all doing time together.

The book *We're All Doing Time,* by Bo Lozoff, published in 1985, revealed a great truth to us. Subtitled "A Guide for Getting Free," it showed that we all are bound in some way, whether we are in prison or out of prison. Lozoff wrote, in a book that was for both inmates and mates outside of prison, "The details differ in a million ways, but at the bottom line, we're all looking for happiness and fulfillment: you, me, the prisoners, guards, wardens, victims, the weak, the strong, the richest of the rich, the poorest of the poor." How are we to serve our time? What is it like to be free?

The Buddhist tradition has brought us to an awareness of the suffering within us and around us. It has made us aware of the need to attend to what is here and now. And that all we do is to be done with compassion. We know that anything done to others is done to ourselves as well. What is done within the prison is done to everyone outside the prison.

How are we to account for the existence of the prison, and for its inordinate use in confining a large portion of the population? The castle stands above us high on the hill. Franz Kafka wished to know the meaning of ultimate things, and in the castle he explored the shadowy interiors of human existence. The world of the castle is cluttered with paper, the white blood of bureaucracy, and with the shuffling of paper, and with the ever-present custodians who shadow K. But nowhere in Kafka's book is it suggested that the castle, simply because it is the seat of authority, merits absolute submission. One can attempt a refusal to submit to the castle.

The castle is a soulless formation, as Irving Howe notes in an introduction to the book, "as if its makers had forever closed their eyes to human needs—or had forgotten the spark that once lighted divinity." The reigning powers are not friendly to humanity. Still, as Howe suggests, K's quest is an authentic yearning for knowledge and truth, even if he lacks the human strength to accomplish the task. There is a kind of heroism in K's quest.

The castle for us is the prison. The prison exists, we rebel against it, and our quest goes on. We ask: Is the prison inevitable? What would a society be like where prisons would not be constructed? We continue to struggle against the prison and all that it stands for, and to struggle against what it does to those inside the prison and to those outside living in the shadows of its foreboding presence.

How are we who remain outside the prison to regard the pain and suffering of those inside? Is not one of the historical reasons for the prison the education of the public? The prison looms before us as a warning that this too could happen to us. In centuries past the torture and execution of the offender in the public square served as warning and lesson. In *The Hidden Injuries of Class,* Richard Sennett and Jonathan Cobb report the observations made by Madame de Sévigné in a letter to a friend in the late seventeenth century. She had witnessed a public hanging of a peasant of the lower realms one morning. Sennett and Cobb write: "It was striking, she records, to see the condemned man trembling during the preliminaries of the execution, when he was only a common peasant. He groaned and wailed incessantly, causing some amusement among the ladies and gentlemen come to see the spectacle; once hoisted up, his body wriggling in the noose, he presented, Madame de Sévigné remarked, a most remarkable sight." The dignity—the life of the peasant—was slightly a matter of concern for the Madame. For other villagers, the spectacle was a clear and gruesome warning beyond amusement. Today, we who are outside the prison watch as others serve their time. We are witnesses to their suffering.

In *Regarding the Pain of Others,* Susan Sontag investigates how we see the suffering of others in the violence brought about by war. Her observations are relevant, in an analogous way, to the sufferings brought about by the domestic arm of the state, the punitive system we know as criminal justice and corrections. Sontag's book is an exploration of, and meditation on, the images—especially the visual images of photography and television journalism—created in the course of war, and the effect these images have on us, the viewing public. We look at the images, with our modern sensibility, with shame as well as shock. What is there we can do? Sontag observes: "Perhaps the only people with the right to look at images of suffering of this extreme order are those who could do something to alleviate it—say, the surgeons at the military hospital where the photograph was taken—or those who could learn from it. The rest of us are voyeurs, whether or not we mean to be." But still, with the images of war, and I would add with the images of the suffering that is caused by crime and punishment, we come closer to identifying with the lives and the sufferings of others.

Our sympathy with the pains and sufferings of others is a complex reaction. Sontag writes: "So far as we feel sympathy, we feel we are not accomplices to what caused the suffering. Our sympathy proclaims our innocence as well as our impotence." Yet sympathy can move us to reflect on the ways in which the institutions and techniques of punishment are constructed and perpetuated.

Rather than becoming immune to the suffering of others because of the constant coverage of war (and crime) in the news, our defenses are broken down, and we open ourselves to the sufferings around us. As Sontag notes: "Let the atrocious images haunt us. Even if they are only tokens, and cannot possibly encompass most of the reality to which they refer, they still perform a vital function. The images say: This is what human beings are capable of doing—may volunteer to do, enthusiastically, self-righteously. Don't forget." We are haunted by the images, and we will not forget.

I have been reading once again the two well-known essays written from prison. One is the essay on civil disobedience by Henry David Thoreau and the other is the letter written by Martin Luther King Jr. from a Birmingham jail. There are many other writings from prison, fortunately, some by those who have served their time and have used their time well to portray and critique imprisonment.

Although Thoreau's time in jail was to last only one night, his reflections on the circumstances and the stay have had a tremendous influence on social and political thinkers such as Tolstoy, Gandhi, and King. It was a night in July of 1846, and Thoreau would later give an account of the revelations that came to him during the night in jail. The night in jail, he wrote, was "like traveling into a far country, such as I had never expected to behold."

In the tradition of the New England transcendentalists, what Thoreau saw in the course of confinement in jail is set against an ideal of what could be. As Thoreau describes his night in jail, physical incarceration is, short of capital punishment, the closest thing to death that the state can perpetuate. The state's response to opposition is tantamount to murder, in Thoreau's imagery. His night in prison is cast as a kind of mythic descent into the depths of hell created by the state. Released in the morning from his dark night in jail, Thoreau imagined a world, perfect and glorious, beyond the prison, "but not yet anywhere seen."

In April of 1963, Martin Luther King Jr. was arrested during a demonstration for civil rights in Birmingham, Alabama. While in jail he wrote his now famous "Letter from a Birmingham Jail," addressed to his fellow clergymen. King was responding to their call a few months earlier for him to allow the battle for integration to continue in the local and federal courts. They had warned that King's nonviolent resistance would incite civil disturbances. King argued in the letter to the ministers that discipleship was at the heart of the African American struggle for freedom, justice, and equality.

Dr. King, near the beginning of the letter, places his call for justice in the tradition of the prophets. He is speaking in the voice of the prophets who left their hometowns and called for justice throughout the land: "I am compelled to carry the gospel of freedom beyond my particular hometown." He continues: "I am cognizant of the interrelatedness of all communities and states. I cannot sit idly by in Atlanta and not be concerned about what happens in Birmingham. Injustice anywhere is a threat to justice everywhere. We are caught in an inescapable network of mutuality, tied in a single garment of destiny. Whatever affects one directly affects all indirectly." He asks his fellow clergymen to be concerned foremost about the conditions that have made the demonstrations necessary.

There are laws that are just, and there are laws that are unjust. The laws that promote segregation are the unjust laws that are, King writes in his letter, "out of harmony with the moral law," laws that must be disobeyed. And, King asks, "Isn't segregation an existential expression of man's tragic separation, an expression of his awful estrangement, his terrible sinfulness? So I can urge men to disobey segregation ordinances because they are morally wrong." The letter from the Birmingham jail ends with words of compassion and hope.

Imprisonment does not usually inspire the visions and hopes of King and Thoreau, shaping our understanding of our time and place in the words written from prison. The prison most often takes us to the place that removes us from our fellow beings. These two courageous instances in our history, Thoreau and King in jail, provide us all with clear and vivid calls for justice, calls that are intimately connected to the institutions and the cultures that would separate us from each another.

I continue to be intrigued by the idea—by the fact—that some people, some of the time, experience a freedom whether they are in prison or out of prison. And that for some—with correct attitude and practice—there is as much freedom in prison as there is outside. Of course, one wonders, if there can be a sense of being free in prison, then what does this say about the world outside? All of our thinking and talking about freedom depends on what we mean by freedom, and what kinds of freedom we are thinking and talking about. Certainly a distinction must be made between the inner being of an individual and the forms of control outside the person.

Thich Nhat Hanh has taught in prisons, and his books have been distributed in prisons throughout the country. One time he visited a maximum-security prison in Maryland, and his teaching there was transcribed and published in a little book titled *Be Free Where You Are*. In a section near the beginning of the book, called "Cultivating Freedom," he writes, "For me, there is no happiness without freedom, and freedom is not given to us by anyone; we have to cultivate it ourselves." He continues:

> You can practice freedom every moment of your daily life. Every step you take can help you reclaim your freedom. Every breath you take can help you develop and cultivate your freedom. When you eat, eat as a free person. When you walk, walk as a free person. When you breathe, breathe as a free person. This is possible anywhere.

The implication from the observations of Thich Nhat Hanh is that we can be *outside* the prison and still not be free. He writes:

> Everyone walks on the Earth, but there are those who walk like slaves, with no freedom at all. They are sucked in by the future or by the past, and they are not capable of dwelling in the here and now, where life is available. If we get caught up in our worries, our despair, our regrets about the past, and our fears of the future in our everyday lives, we are not free people. We are not capable of establishing ourselves in the here and now.

And one person, Thich Nhat Hanh adds, can affect the world. "One person sitting, walking, eating, and breathing as a free person can make an impact on the whole environment around him."

Where does this journey into the nature of the prison, and the nature of our human being, take us? I have learned that bondage occurs from the ignorance of not being aware, of not being mindful. Freedom becomes possible when we are not bound by preconceived thoughts, when we attend to the concrete moments of the present. But even freedom as a thought, as Steve Hagen notes in *Buddhism Is Not What You Think,* can be a source of bondage. Hagen writes: "Thus we have a difficult time truly tasting freedom because the more we think about it, chase after it, and try to reach it or acquire it, the more we only thwart our efforts by throwing ourselves back into bondage. In trying to obtain freedom, we perpetuate our confusion and suffering." Any thought pursued beyond its rightful moment leads to bondage. We can be bound even by thoughts of liberation.

Seeing deeply into the reality of the situation at hand, into what is happening, is liberation in itself. When we see clearly that separation between self and other is a false perception of reality, we understand that harm done to one is harm done to all, and that the care taken with one is care given to everyone. It is all of one fabric, this web of life. We are not so much our brother's and sister's keeper as we are one with our brother and sister. Hagen asks: "What if you saw that 'I'm here and he's there' isn't a full explanation of reality? What if you saw that injury to your brother is injury to yourself?" There is no world out there that is separate from the one in here. When you are in prison, I am in prison also.

Prisons will exist as long as we humans are not released from the bondages that keep us all enslaved. The prison represents—and is a reflection of—the bonds we place on ourselves in our everyday lives. To be aware of what is actually happening here and now is to know and experience reality. Only when we are in tune with—in harmony with—the impermanent, ever-changing reality, will we begin to break down the walls of the prison. Abolition of the prison is an integral and seamless part of enlightenment and human liberation.

There is a loud background noise, a humming in the universe. The war in Iraq continues, and dominates the daily news. The death toll increases daily. For the American forces, thousands of men and women have died. Tens of thousands of Iraqi citizens have been killed. The people of other nations have increasingly become alienated by American arrogance and violence. Max Rigert, a retired minister in Delavan, writes to the local newspaper quoting the words of the writer, an ancient philosopher, of Ecclesiastes: "The quiet words of the wise are more to be heeded than the shouting of a ruler among fools, for wisdom is better than weapons of war." The minister exclaims, "War is still not the answer!"

Two people, a man and a woman, meet in a café. They have been brought together this late afternoon to introduce themselves to each other and to plan the exhibit they will have together in the café later in the year. They select a table in the corner. She orders a latte and he orders the black coffee of the day. For half an hour they talk about their interests in photography and the work they have done. She unpacks her laptop and shows him a selection of her color photographs. He has brought a few small prints of his black and white photographs to show her. They gradually turn to details, vaguely conveyed, about their families and the places they have lived. She tells him that she once knew someone with his last name. And recalls the place and years, the college years, of this knowing. Silence for some time, and then he asks, hesitantly, if her name might have been—such and such—fifty years ago. They, now, across the table, see in the face of the other the one they had known years ago, and known in some romance. He had wondered for fifty years where she had gone, what had become of her, and might his life have been otherwise. Meetings can take you to unexpected places. Especially in cafés, in the afternoon, late in one's life.

We have returned from a two-week trip to Norway, to the country of mountains, fjords, rocks, rain, and the days with the sun not setting until midnight. I took with me, for occasional reading, a short biography of J. M. W. Turner and a small volume of Zen poetry. The fifteenth-century poet Ikkyu Sojun reminded me:

Look very closely:
Only impermanence lasts.
This floating world, too, will pass.

And this about ourselves:

Like vanishing dew,
a passing apparition or the sudden flash
of lightning—already gone—
thus should one regard one's self.

We roomed, graciously hosted by relatives, at the edge of the North Sea.

Together in the car, Solveig and I and cousins traveled under the sea, through the long tunnel, every time we went from the island into the city of Stavanger. Lapwings fed in the open fields, the birds gathered on these midsummer days for their flight south. Cars sped toward the ferry that will take travelers north to Bergen. Sheep grazed on the hillsides of ancient farms marked by stone fences. In the park Bjersted, we sat in the Stavanger Konserthus listening to a jazz group playing and singing a repertoire of American songs, with lovely renditions that included the songs of Cole Porter and George Gershwin. The program moved from "I've Got You Under My Skin," to "Let's Fall in Love," to "The Nearness of You." The concert concluded with a welcomed version of "I'm Beginning to See the Light." Before returning to the island, we walked through the town and circled the harbor, now holding three large cruise ships that have brought other tourists from afar.

Wherever and whenever Turner would travel, he filled sketchbooks with detailed observations that might serve him when he returned to paint in his studio. Sketching was a way of paying close attention to the natural world, this floating world, until the end of his days. In a gallery in Stavanger, I carefully studied a painting by the Norwegian artist, now passed away, Harald Kihle. It is of a painter traveling through the Norwegian countryside on his Fjord horse, sketchbook and tools at his side. I think of myself on this trip to Norway, carrying my camera loaded with black and white film, as such an artist, sketching—or about to sketch—the wonders (large and small) all around me. Romantically imagined, I'm a sketcher, a Turner in my time.

On the wall in the hallway of my house is a hand-tinted print of a stranger entering a village in the west of Ireland. The print is by Jack B. Yeats, and it is titled "The Village." I purchased it some years ago in a shop in the tower that once was the home of W. B. Yeats, his brother. Horses and carts and villagers are on the street lined with thatched cottages. Children can be seen in the distance running from the school at the edge of the village. Something is about to happen as a stranger, knapsack on his back and walking stick in hand, enters town.

Earlier in the summer, I listened carefully to a poem being read early in the morning on the radio, titled "Who Was That Man?" by Paul Bussan:

> I love those movies
> About a stranger
> Who rides into town
> On the back of a horse,
> And proceeds to start
> A chain of events

That makes each person
Take stock of their lives,
So that after he's gone
Everyone's better
Or worse for the wear
Than they were
Before he arrived.

I continue to entertain what may come without expectation. How this life happens beyond all the plans in the world.

The myth I keep returning to, the myth of Sisyphus. It first came to me as a young man, sitting beside an outdoor swimming pool in the hot afternoon sun in a southern town. A point in my life when the absurd seemed to be as real as anything else. I sat reading, with feet dangling in the cool water, *The Myth of Sisyphus* by Albert Camus.

In the Greek myth, Sisyphus repeatedly rolls a huge stone up a hill only to have it roll down again. Sisyphus has been condemned for cheating death, escaping from Hades, and returning home. A trickster, a popular figure in folklore. Camus writes, "His scorn of the gods, his hatred of death, and his passion for life won him that unspeakable penalty in which the whole being is exerted toward accomplishing nothing." This is the price, Camus adds, that must be paid for the passions of this earth.

I continue to read, over the course of the stages of my life, the concluding paragraph on Sisyphus.

> I leave Sisyphus at the foot of the mountain! One always finds one's burden again. But Sisyphus teaches the higher fidelity that negates the gods and raises rocks. He too concludes that all is

> well. This universe henceforth without a master seems to him neither sterile nor futile. Each atom of that stone, each mineral flake of the night-filled mountain, in itself forms a world. The struggle itself toward the heights is enough to fill a man's heart. One must imagine Sisyphus happy.

A negation of the gods, entertainment of the unknown, hatred of death, and happiness in the struggle of being human. A myth to live by.

The Dalai Lama begins his book *The Art of Happiness* with these statements:

> I believe that the very purpose of life is to seek happiness. That is clear. Whether one believes in religion or not, whether one believes in this religion or that religion, we are all seeking something better in life. So, I think the very motion of our life is towards happiness.

To seek happiness is our goal, he says, whether one believes in religion or not.

The Dalai Lama continues by distinguishing between two levels of spirituality, one having to do with religious beliefs and the other beyond any such beliefs. He writes, "I believe that each individual should embark upon a spiritual path that is best suited to his or her mental disposition, natural inclination, temperament, belief, family, and cultural background." All religions, he adds, "are designed to make the individual a happier person, and the world a better place." And "religion should be a remedy to help reduce the conflict and suffering in the world, not another source of conflict." But it is "basic spirituality"—with the human qualities of goodness, kindness, compassion, and caring—that is to be cultivated by the world's population. He suggests that four-fifths of the world's population consists of non-believers in a religion, that individual spirituality is their religious path.

The practice for a spiritual life, as found particularly in Buddhism, is the practice of mindfulness. The Dalai Lama puts it this way:

> So, engaging in training or a method of bringing about inner discipline within one's mind is the essence of a religious life, an inner discipline that has the purpose of cultivating these positive mental states. Thus, whether one leads a spiritual life depends on whether one has been successful in bringing about that disciplined, tamed state of mind and translating that state of mind into one's daily actions.

Happiness is built on a foundation of a calm and stable mind, on a "stillness of the mind."

The ultimate purpose of spiritual practice is the development of compassion. When asked to define compassion, the Dalai Lama replied: "Compassion can be roughly defined in terms of a state of mind that is nonviolent, non-harming, and non-aggressive. It is a mental attitude based on the wish for others to be free of their suffering and is associated with a sense of commitment, responsibility, and respect towards the other." Genuine compassion, he notes, is based on the awareness that all people have an innate desire and need to be happy, to overcome their suffering. There is a sense of commonality with all living things, and the intentional cultivation of patience and tolerance. Compassion, tolerance, patience, and caring for the welfare of others are the basis of the religious life.

These core values are to be observed and practiced, concludes the Dalai Lama, no matter what particular religion one may follow. Respect for other religions and compassion for the spirituality of our fellow human beings are integral to a spiritual life. The primary aim of all religions—and spirituality in general—is to help us become better human beings.

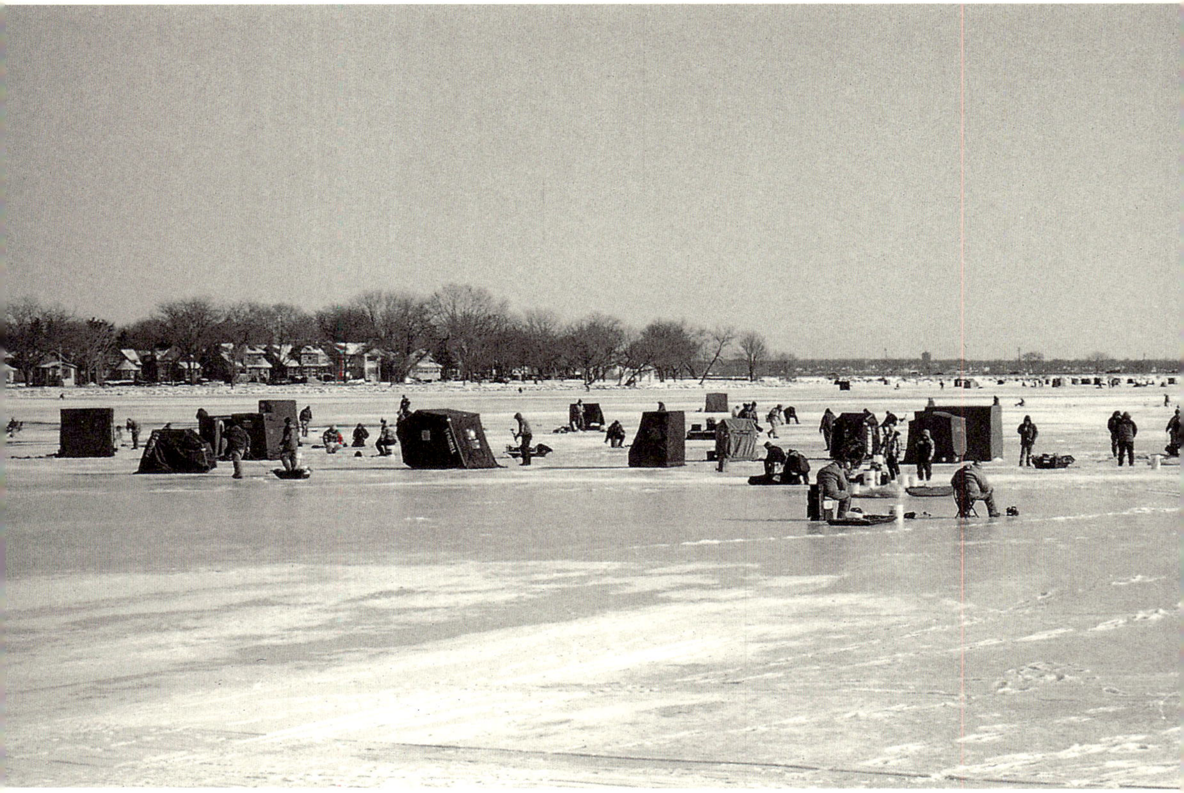

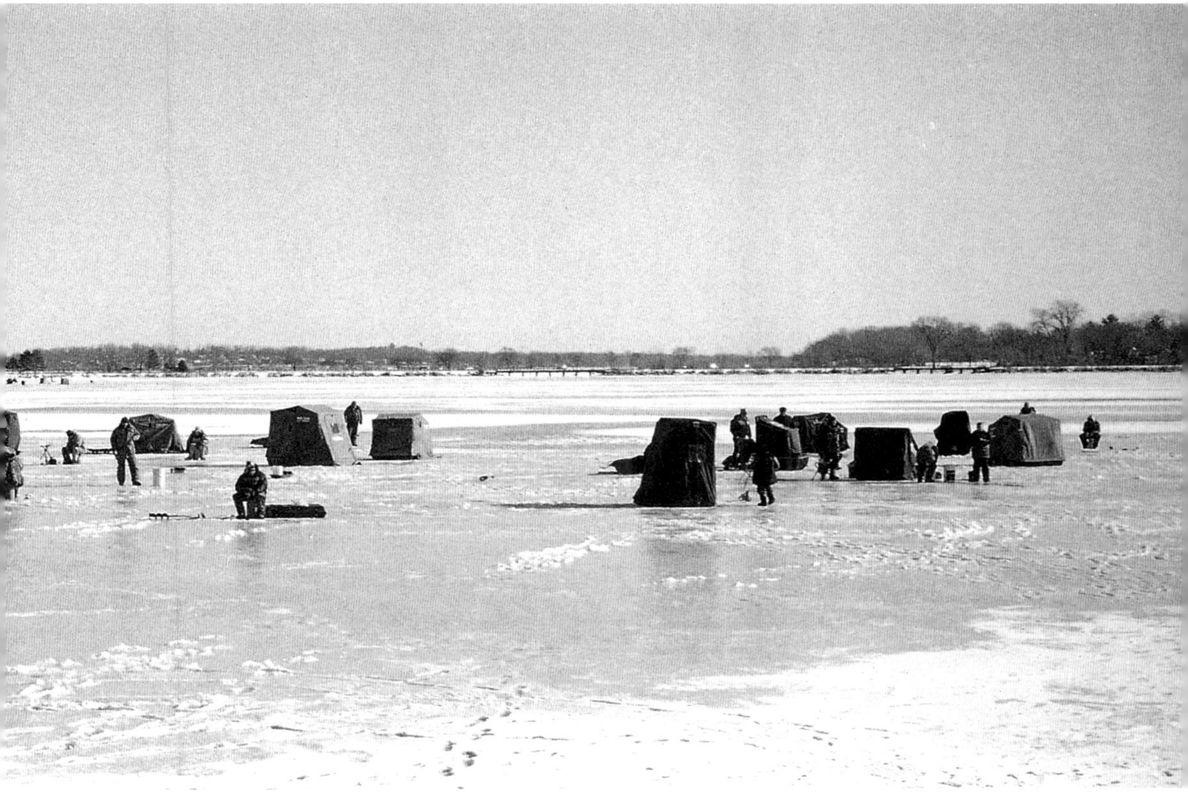

Part IV

What We Need Is Here

And we pray, not
for new earth or heaven, but to be
quiet in heart, and in eye
clear. What we need is here.
Wendell Berry, "The Wild Geese"

We come finally to this, to where we have been all along without knowing it, to the precious present suspended between a past of time no longer existing and a future yet to be. Our only true time is the brief moment consciously experienced between forgetting and anticipating. A borderline marks our world of reality.

I have taken to the road once again to see the places where earth meets the sky. The long horizon—spreading all around—is the geographical line that separates earth and sky. The equivalent of time present experienced between the remembered past and the imagined future.

These roads and byways that I travel, often so close to home, provide me with all the adventure I need or desire. I know a great joy when I see the earth meeting the sky. In an instance, ever so briefly, I experience the immediacy of the here and now. This is my religion as a traveler on the road. The holy experienced daily, moment by moment.

That which we imagine as the divine can be seen spreading across the land. A visual image—seen directly or viewed in a painting or a photograph—transports us into realms most sacred. Henry David Thoreau reminds us, "The eyes were not made for such groveling uses as they are now put to and worn out by, but to behold beauty now visible. May we not see God?" With eyes wide open, new sights and visions come into view. Here on earth, all that we can imagine as the divine is before us as we travel.

And travel need not be to extraordinary places with exotic landscapes. No place special is necessary to see anew. Wendell Berry ends his poem "The Wild Geese" with the line, "What we need is here." A Unitarian minister, with this poem in mind, writes: "What we need is here—in the gardens we tend, on the morning walks we take, in the faces of our children, in the stars above our heads at night. It is even in the busy city streets, and still in the world of nature, changed though it may be from the time of Emerson and Thoreau and their contemporaries. We have only to be patient, to take the time to slow down and look around us to understand that it is true."

This natural world is the answer to all our yearnings and desires. Solveig has shown me an obituary of an ancestor in Norway where mention is made of the singing of a hymn at the funeral service, "Deilig er jorden," a hymn that translates into English as "Beauty of the Earth." This is the hymn that in this country is titled "Beautiful Savior." In the natural world, on another shore overlooking the North Sea, in the meeting of earth and sky, is a spirituality that is beyond personification in a deity. All for the eye to see. "I once was lost, but now am found, was blind, but now I see." Good words for the many roads we travel.

My father died at the age of sixty-nine—in the year 1969. A few short years ago I reached the age he died. Each year since I have known that I live a bonus. Now I live two lives as one.

Maybe this accounts in part for my being drawn to the far horizon, beyond the established borders and boundaries, to where the earth meets the sky. In his autobiographical essay "Something to Write Home About," Irish poet Seamus Heaney describes the god the Romans called Terminus, the god of boundaries. An image of the god was kept in the Temple of Jupiter on Capitol Hill. The sky was visible through an opening in the roof, "as if to say that a god of the boundaries and the borders of the earth needed to have access to the boundless, the whole unlimited height and width and depth of the heavens themselves." Even with boundaries and borders, both gods and humans need to feel that they are unbounded. Heaney writes that a good poem "allows you to have your feet on the ground and your head in the air simultaneously."

And for me, as I search the horizons, a good photograph allows me to be grounded on earth but to be in the sky at the same time. Earthbound and open to the spaces beyond. Here and elsewhere in the same moment, in the simple release of the shutter.

A few slow and graceful chords, and the well-known voice comes through loud and clear: "I live the kind of life most men dream of. I make my living writing songs and singing them." As Merle Haggard sings his country song, I'm about to cross the last glacial moraine a few miles west of town. Another borderline, this one marking the divide between the glaciation that ended ten thousand years ago and the old landscape we call "driftless." I'm completely at home today traveling the hills and valleys of this driftless land.

A deeply dissected terrain, well drained by streams and tributaries, outcropped with Paleozoic bedrock, this is the driftless country that I travel. A landscape shaped by a great glacier advancing and receding over a period of a hundred thousand years. The moraine ridges were deposited and shaped as the glacier finally melted. These moraines sit on top of an even earlier glaciation. Layer upon layer, millions and billions of years, and I ride along the ridges now, an ephemeral being going nowhere and everywhere all at once. I am struck by the reality of impermanence, the earth's and mine.

I think of William James and his phenomenology of time as I cross the Johnston Moraine on Mineral Point Road. Time as we humans experience it. Our apprehension of time consists of four parts: the obvious past, the specious or recent past, the real present, and the future. Three of these four are nonentities, the only real time is the present moment, the moment that escapes us as we entertain the recent past, the passing of time as we speak, and the time expected soon. Time in human perception is almost completely on the fringes of the real present. These fringes, fore and aft, vary in length of time from a few seconds to not more than a minute. Truly, our lives are lived on the fringes of reality.

And the same is true with our perception of space. Space for James is not an abstraction, but is a "simple, specific quality of retinal or cutaneous sensation." What I see in the distance as I view the landscape before me in the viewfinder of my camera is a space perceived by my senses, physiologically and psychologically. This is my space, humanly perceived. Framed within the camera's viewfinder is a long horizon with sky above and earth below. And the space above and below the horizon, devoted to earth and sky, is similar to my sense of the duration of time. The horizon is the real present, and radiating above and below are a past and a future. But still of considerable importance. These are the times and the places I know so well.

Three sharp strokes of the bow across the strings of the violin, and then the haunting voice of George Jones singing about cowboys not crying and heroes not dying. About good always winning again and again, followed by words of love and loss. The chorus line could break your heart: "But here in the real world, it's not easy at all." The bright morning sun is moving across the midwinter sky. Lucky I am to be on the road, again, heading into the glacial land east and south of town. My own mantra for the day is the question asked by St. Francis Assisi, "Lord, what shall I do with my life?"

I drive these roads randomly, but not aimlessly, through the glacial landscape. Here now—the drumlins, the kames, the kettles and the moraines, eskers, lakes and streams formed by the fluctuating lobes of the last Wisconsin glacier. This land is still recovering from the melting glacier of ten thousand years ago. Consulting a guide for the Ice Age Trail, I am informed: "As the streams slowly wash away kames, eskers, and moraines, and as marshes, bogs, and lakes fill with sediment and organic debris, the young landscape will become like the older glacial landscape which lies between the Driftless area and the terminal moraines of the most recent glaciation." With time, as with the transience of all nature animate and inanimate, even this landscape will change. Here and there, for a while, will be a remnant hill and valley. All that you and I know now will someday be long ago.

Traveling over the land, stopping occasionally to photograph, I adjust my perception of time and space. Gradually, I become more aware and attuned to the land and my movement through it. In the most heightened moments, I identify with the land, with the meeting of the land and the sky, and I become one with heaven and earth. I then pull to the side of the road, take my camera in hand, carefully compose in the viewfinder, and release the shutter. From the developed film, eventually, some other day, I will select negatives for the printing of photographs. These photographs will be the artifacts of the moments I have come close to being a conscious part of heaven and earth.

The photograph is the byproduct of the union of the seer and the seen. Proof—evidence—that such union once existed. My meditation has been at the side of the road, camera raised to the horizon. I have been inspired, intellectually and spiritually, by the example of the artist Frederick Franck in his "seeing/drawing." In his book of handwritten text and drawings, titled *The Zen of Seeing,* he compares drawing to zazen, sitting meditation:

> Zazen may be the discipline of Zen, but I for one, am not good at sitting still for long in the lotus position. I believe that in SEEING/DRAWING there is a way of awakening the "Third Eye," of focusing attention until it turns into contemplation, and from there to the inexpressible fullness, where the split between the seer and what is seen is obliterated. Eye, ear, hand become one with what is seen and drawn, things are seen as they are—in their "isness." Seeing things thus, I KNOW WHO I AM!

This eye that sees—focused on the meeting of earth and sky—is the I that is experiencing itself. I am realizing, experiencing, my being an integral part of nature, part of all that is. Traveling toward the horizon is a spiritual practice, a religious experience in the ordinary living of the day.

The snow came in the night and continues throughout the day. Tomorrow or the next day, or the next, when the sun will shine again, I will drive the country roads and watch the trees climb the snowy hills. And as night comes, I may walk on the snow-covered street near home. Much like the walker at the end of David Budbill's poem "Winter: Tonight: Sunset" which I have just read.

> I pause in this moment at the beginning of my old age
> and I say a prayer of gratitude for getting to this evening
>
> a prayer for being here, today, now, alive
> in this life, in this evening, under this sky.

What comes to mind on a winter's day is different from thoughts, my random thoughts, on a day in summer. The reading of Robert Creeley's last words on poetry and aging prompts my thoughts at the moment. I am keenly attuned to his lament in his book *On Earth,* "the end seems never far enough away."

A short time ago I asked these questions in my notebook: "What do I now care about—enough to photograph?" "What at this time in my life is of interest to me?" "What meaning may I give to my life as I grow older?" Obviously, expressions of uneasiness, and a need for spiritual sustenance, for what is conventionally called salvation. Questions of faith, of what may inspire careful attention to the matters of everyday life.

"Fancy" is the term Creeley and other poets give to the desire to live with caring heart and mind. To live with the sense that life is worth living. (Camus wrote that the one truly serious philosophical problem is that of suicide.) A confidence that our conscious world has not yet fallen from us in our aging and our infirmity. This is the fancy that Walt Whitman celebrated in his poem "Good-Bye My Fancy," ending the poem with the line, "Good-bye—and hail! My Fancy."

You know when this fancy fades in your life. Maybe even in the course of a day. Faith may be little more than maintaining, or regaining, fancy. And you hope that fancy will not leave you before you reach the end of your days. Or, it may be that the letting go of fancy is what makes death bearable in the end. Fancy may be life itself. Without fancy, what is life?

Creeley's last book contains poems about the diminishment of fancy in old age. In the poem "Sad Walk," he compares walks when young to walks when old.

> But now it's just a sad walk
> to an empty park,
> to sit down and wait, wait to get out.

But the poet, in the poem simply titled "Here," ends with a fancy that has been in his life from the beginning: "Keep looking, keep looking / Keep looking." I, too, will keep looking as long as grace grants the desire to look. I will look this day, earth covered with snow, at the bare limbs and branches of the trees elegantly reaching into the sky.

The lyrics to "Who Will Watch the Home Place?" are being sung by Laurie Lewis on the car radio as I back out of the garage and begin my trip down to the farm. Sixty miles south through the drumlin ridges and kettle moraines. I welcome the sparkling snow on the ground as I pull into the driveway at the farm. Last fall's fallen Chinese elms are piled high in the yard, ready for the taking by neighbors for their stoves and fireplaces. Black Angus heifers, bred for spring calving, gather around the steaming water tank. The three horses move along the fence line, forming constellations as if floating in a powdery white sky. I am at home, the original one, once again.

On my way down to the farm, Garrison Keillor, on his daily *Writer's Almanac,* reads a poem by one of my favorite poets, William Stafford. The poem, "What's in My Journal," reminds me of these field notes that I am making. Notes kept as some sort of religious act, a living with intention, and a faith that life is worth living. Notes in a journal not clearly decipherable, but plumbing the deep waters of being human.

> Odd things, like a button drawer. Mean
> things, fishhooks, barbs in your hand.
> But marbles too. A genius for being agreeable.
> Junkyard crucifixes, voluptuous
> discards. Space for knickknacks, and for
> Alaska. Evidence to hang me, or to beatify.
> Clues that lead nowhere, that never connected
> anyway. Deliberate obfuscation, the kind
> that takes genius. Chasms in character.
> Loud omissions. Mornings that yawn above
> a new grave. Pages you know exist

but you can't find them. Someone's terribly
inevitable life story, maybe mine.

The home place is a poem with infinite lines. I add new verses to the poem each time I return. This time Laurie Lewis is singing questions about who will watch the home place. Asking, finally, "Who will fill my empty place / When I'm gone from here?" I'll stop in Whitewater for the late luncheon special at The Sweet Spot, lingering for a while along the railroad tracks in the town.

We humans, with a few changes in DNA, are best distinguished in the kingdom by the evolution of consciousness. A consciousness that meanders like a stream without course or design. The mind playfully dancing, much like snow falling and swirling this beautiful morning. But the human mind can deliberately focus, look outward, or peer into itself. In Buddhism, this mental discipline is called *bhavana,* a mental training usually translated as "meditation." Deliberately, introspectively and mindfully, probing deeply into the nature of a chosen object. This human capacity for disciplined consciousness is described as an empirical science in the Dalai Lama's *The Universe in a Single Atom,* subtitled "The Convergence of Science and Spirituality."

The Dalai Lama argues that one's consciousness, as well as human consciousness in general, can be altered, transformed, by the disciplined mind. Changing how and what we think can change our attitudes and emotions. This means, especially, "that the cultivation of loving-kindness can over a period diminish the force of hate in the mind." Within our minds is the natural ability to transform consciousness and, subsequently, behavior. In the cultivation of mindfulness, we can develop equanimity and become more compassionate in our everyday lives. "The theory of Buddha nature—the notion that the natural capacity of perfectibility lies within each of us—is a deeply and continually inspiring concept."

Focusing the camera on a distant landscape brings the consciousness of the photographer to a point of concentration. The mind becomes focused, just as the camera is focused, and stream of consciousness slows down as mind's eye sees clearly. Taking notes in the field concentrates the mind as well. Photography and note-taking are meditative practices. Consciousness is transformed on the roads traveled close to home.

It is below zero this bright morning. The counter-intuitive headline proclaims that global warming is "unequivocal." The international network of climate scientists has concluded that the world is in the midst of climbing temperatures, shifts in weather patterns, and rising seas—as a result of the buildup of gases in the atmosphere caused by human activity. Hopefully, the harmful consequences can be blunted by human intervention. In the meantime, carbon dioxide is increasing in the atmosphere, temperatures have risen, and there is evidence of an increase in hurricanes. By the end of this century, as illustrated on a world map, the surface temperature will have turned from the cool blues and greens to the hot oranges and reds. From my personal point of view, I am sorry that I will be missing the world no matter what color the map.

A friend died last week, leaving this world of what we know as life. "Passed," as we say, losing body and earthbound spirit. During the Second World War, serving as a navigator on a B-11, he and his crew were blown out of the sky. He suffered physical and mental injuries for the rest of his life, as noted in the obituary he had written earlier for the local newspaper. At his memorial service, "A Celebration of Life," a poem he had written was read just before a closing song to sing sorrows away.

Ed had prepared the final service, noting the order of service, the hymns and special music, the readings, and some of the speakers. For the opening words, he had selected something I had written some years ago. A call to those gathered for various occasions, I titled the piece "We Gather Together." I'll enter it here for anyone to use, in times of joy and times of sorrow.

We who dwell on a vast continent
That stretches into the universe
Gather together.

In times of joy and times of sorrow,
In the changing of the seasons,
We gather together—to care for one another,
To live our lives with purpose,
To find wisdom, compassion, and strength,
And to know the oneness of all things.

Let us listen.
Let us follow the rhythm of life's mystery,
And join in its dance.

Formed each moment by the way we live,
We give thanks
For this life together.

The unstable and tragic experience of the first movements of Gustav Mahler's Fifth Symphony will give way to the more optimistic conclusion. An emotional release will be heard in the fourth movement, the famous Adagietto. There will be a celebration of sorts in the Finale. All on this cold and happy morning, sky filled with falling snow.

In composing this and the other symphonies, Mahler was expressing, and seeking solutions to, the existential problems of his personal and spiritual life. Is this not the source of all composing, musical and otherwise? And in the making of a photograph, or in the few notes made in the journal, someone is seeking a resolution to the human condition. In religious terms, a redemption—a response to something imagined in the Garden a long time ago.

My friend Dennis, with compassion and humor, refers again to my plodding in the existential realm. To my search for purpose and meaning when there seems to be no answer outside this earthly existence. Although my course will not be easily diverted, ingrained as it is in my nature, Dennis offers another course, firmly grounded, that guides his life. It is the tradition of Ruskin, Marx, Kropotkin, Proudhon, Shelley, Blake, and Ginsberg—that this life is the eternal life when lived fully.

Dennis writes, "It is impossible to possess divine (eternal) life without creating and living within a sustaining community." What may be known as God, the divine, is found in a community of people helping each other. "The practice of mutual aid is God, to be brothers and sisters is God." Life is worth living, we are saved, and all the manifestations of death—fear, anxiety, despair, and depression—are dissipated when we are in communion with others. In community, eternity is created in this life.

The Finale in Mahler's Fifth Symphony follows the doubt and suffering of the earlier movements. Finally there is a hard-won, somewhat optimistic, albeit ambiguous, conclusion. This is cause for celebration.

Sisyphus's rock will inevitably roll back to the bottom of the hill. And Sisyphus will labor again, this time with the help of others, in the struggle to be human. To know the eternal here on earth.

You wake up in the morning—following the all-night snow-storm, the storm that began early in the afternoon the day before. How many times has this happened in your lifetime? Maybe four or five times that you can remember as distinct and separate times. You remember the storm when you were a child, in the mid-1940s, that left the family snow-bound for days. Snowed in until one day the orange snowplow made its way down the country road piling the drifts higher than the cars and trucks that now could pass.

Nineteenth-century nature writer John Burroughs, in his essay "A Snow-Storm," eloquently described such a storm to thousands of schoolchildren in one-room schoolhouses across the country. In three pages, he took students through a snowstorm, from the approaching storm to the end of the storm hours later. The preparations of the storm are gentle and quiet: "A marked hush pervades both the earth and the sky. There is no uproar. no clashing of arms, no blowing of wind-trumpets. These soft, feathery, exquisite crystals are formed as if in the silence and privacy of the inner cloud-chambers. Rude winds would break the spell and mar the process." And after many hours, at the end of the storm, in Burroughs's poetic words, this is what we behold:

> How deeply, and warmly, and spotlessly Earth's nakedness is clothed! It is a veritable fleece, beneath which the shivering earth is restored to warmth. And how, like a fleece of wool, it rounds and fills out the landscape, and makes the leanest and most angular places appear smooth and beautiful!

"The Snow-Storm," by John Burroughs, is found in *Sheldons' Modern School Fifth Reader,* the reader that my grandfather used in grade school in the 1880s. My grandfather, Will Holloway, and his fellow students in the seventh and eighth grades read the essays and poems of Alfred Tennyson, Joseph Addison, Thomas Carlyle, William Shakespeare, Henry Wadsworth Longfellow, Charles Dickens, Washington Irving, Elizabeth Browning, Percy Shelley, John Greenleaf Whittier, George Eliot, and many other writers known and unknown to us today. The Preface to the *Fifth Reader* notes, "It is believed that these selections possess literary merit, making them worthy of close study as specimens of good English." Each reading is to be carefully studied by the students for learning new words, but foremost is the preparation for reading aloud to each other in the schoolroom.

In my own education, in the one-room Dunham School in the 1940s, I cannot recall such close attention being given to the study of literary works. The *Fifth Reader* of the 1880s advised students to read slowly and to read with a thoughtful mind.

> Accustom yourself then to thinking. Set yourself to understand whatever you see or read. To run through a book is not a difficult task, nor is it a very profitable one. To understand a few pages only is far better than to read the whole, where mere reading is

> the only object. If the work does not set you to thinking, either you or the author must be very deficient.

And to my surprise now, as I continue to read my grandfather's *Fifth Reader,* is the preparation for reading out loud: "Good reading, as a vocal exercise, is the proper expression of thought and feeling in speaking tones." Image the students preparing to read the assigned literary work aloud. "Each piece must be carefully studied, and its nature and design be fully comprehended, before it can be read either by teacher or pupil." The opening pages of the *Fifth Reader* provided the instructions and illustrations for good and proper reading, devoting attention to "Pronunciation, Articulation, Cognate Sounds, Accent, Inflection, Emphasis, Modulation (including Pitch, Compass, Quantity, Quality, and Force), Pauses, and Word Study," with special instructions for the reading of poetry. "The first requisite for such reading," the *Fifth Reader* notes, "is that the thoughts and feelings to be expressed shall be thoroughly understood and felt by the reader. Both the intellect and the emotions are concerned in all true reading." Above a sentence in his reading of a selection on the passing of "The American Indian," my grandfather has written in pencil, indicating the emotional tone of the passage, "showing passion."

In such reading of literary works, students were learning the practices that would serve them well as readers, and as writers, for the rest of their lives. My mother's letters to me, after I left home, the letters I now treasure and save, are in the tradition of her father's (my grandfather's) preparation for good reading. Reading and writing are of one fabric, of the same collection of skills, thought, discipline, and passion. A voice is heard in the writer's head—and heart—that was honed and heard in the one-room schoolhouse not far from where I am writing these notes on another winter's day.

Late in his life, a clergyman asked John Burroughs if he would come and talk to the congregation on "The Gospel of Nature." I don't know if Burroughs ever delivered his thoughts on the topic, but he was so inspired by the request that he wrote an essay on the gospel of Nature. He began the essay with questions to himself:

> What had all my many years of journeyings to Nature yielded me that would supplement or reinforce the gospel he was preaching? Had the birds taught me any valuable lessons? Had the four-footed beasts? Had the insects? Had the flowers, the trees, the soil, the coming and going of the seasons? Had I really found sermons in stones, books in running brooks, and good in everything? Had the lilies of the field, that neither toil nor spin, and yet are more royally clad than Solomon in all his glory, helped me in any way to clothe myself with humility, with justice, with truthfulness?

Burroughs recognized the difficulty of responding to a question about the religious nature of Nature. Yet he was convinced of a strength and solace given to him by all his years of exploring the natural world: "Natural influences work indirectly as well as directly; they work upon the subconscious, as well as upon the conscious, self. That I am a saner, healthier, more contented man, with truer standards of life, for all my loiterings in the fields and woods, I am fully convinced."

The forms and creeds of religions change, "but the sentiment of religion—the wonder and reverence and love we feel in the presence of the inscrutable universe—persists." For Burroughs, as with his spiritual predecessors—Wordsworth, Emerson, Thoreau, and Whitman—the word Nature is very much as others have used the word God. "I suppose, back of it all we mean the power that is everywhere present and active, and in whose lap the visible universe is held and nourished." The terrestrial and the celestial are one, as are time and eternity, mind and matter, life and death, and there is nothing that is not inherent in Nature. "Then we no longer look for or expect a far-off, unknown God." To know this, to experience it in your life, you must be open to the natural world, to the lives of birds, for example, taking heed of the natural world with interest and affection, associating with it daily, spring, summer, winter, and fall.

But note, from a human stance: "Nature is not benevolent; Nature is just, gives pound for pound, measure for measure, makes no exceptions, never tempers her decrees with mercy, or winks at any infringement of her laws." This is a hard gospel, Burroughs tells us, "but rocks are hard too, yet they form the foundations of the hills." We humans introduce benevolence, mercy, and altruism into the world, and we pay the price in what we reap in the lives we create here on earth.

And this form we know as the human being, Burroughs writes in his gospel of Nature, is not the aim and end of creation. The vast universe—all of Nature—is of one piece, and there is life beyond this world, beyond this planet that will someday come to an end. "What then? The game will be played over and over again in other worlds, without approaching any nearer the final end than we are now. There is no final end, as there was no absolute beginning, and can be none with the infinite." Such is the higher God, the infinite one.

Many times, just when you are about to be content, to settle into a perspective that will furnish all the grounding you think you will need, you are drawn to disorder and chaos again. Maybe it's because you need to be unsettled, to know and to feel what it is to be human. Coming in on the radio this morning is George Strait singing "Let's Fall to Pieces Together." I welcome the song on Bill Malone's program of classic country music.

And the morning newspaper has a review of the last of Tom Stoppard's trilogy "The Coast of Utopia." About the lives of nineteenth-century Russian revolutionaries, the reviewer notes that it is a play for all of us. Life is filled with upheavals and unpredictability, and is a mess "as consuming and capricious as the ocean storms that are evoked so exquisitely by the production's technical wizards." Stoppard has the central figure of the play, the political theorist Alexander Herzen, say in the concluding scene: "It takes wit and courage to make our way while our way is making us, with no consolation to count on but art and the summer lightning of personal happiness."

Such is the impossibility, finally, of imposing order on the amazing and wonderful contradictions of human existence, on matters both domestic and transcendent. Oh, I couldn't be happier today.

Joining in life's dance. Piano arpeggios accompany the snow falling and blowing in the wind on the way to earth. The dance that is Shiva's dance. What John Burroughs meant as Nature. All of our human speculations about the meaning of life—and about our place in a larger universe—are manifestations of our divinity, of our relation to something beyond us and ultimately unknowable. Something beyond the perishable cosmos. The unmanifest as imagined in *The Bhagavad Gita* and the *Upanishads*. We are of that, *tat tvam asi*.

Charles Darwin began his observations as a naturalist on the voyage of the *Beagle*. With the publication of *The Origin of Species* years later in 1859, he altered drastically the dogmatic Christian belief on the creation and meaning of human life. While writing the book, his young daughter died, causing Darwin to abandon his remaining vestiges of Christian faith. In an essay titled "Rewriting Nature," Adam Gopnik describes concisely Darwin's final stoical stance: "He believed that the contemplation of the immensity of time, and the repertory of feelings, was all that was left to us. There was no inherent meaning in Annie's dying at ten, except the recognition that mortality was the rule of existence; serenity could be found only in the contemplation of the vast indifference of the universe." But still, Darwin would write, given the indifference of nature, that "there is grandeur in this view of life."

Time is both human and cosmic, within human grasp and beyond our imagination. There is no bridge between the two times. I like Gopnik's description of Darwin's resolution of the two times:

> The tragedy of life is not that there is no God but that the generations through which it progresses are too tiny to count very much. There isn't a special providence in the fall of a sparrow, but try telling that to the sparrows. The human challenge that Darwin felt, and that his work still presents, is to see both times truly—not to attempt to humanize deep time, or to dismiss quick time, but to make enough of both without overlooking either.

As with Darwin, what is left for us is the chance to respect, to love, and to revere nature. Even if nature cannot care for us, we can care for one another. Our lives are shaped by nature, but this leaves much room for loving this life and each other within our own time. For Darwin, says Gopnik: "He thought that he had found the secret of life. But he knew that nothing could solve the problems of living. That takes all the time we have."

In the meantime, Shiva dances in the burning circle of fire. (The bronze statue of Shiva rests on the high shelf of my bookcase.) The cosmic energy of Shiva will destroy the present world and create the new world, as a repeating cycle. We are someplace in that energy, in the destruction and the creation. Dancing is my meditation this winter day, knowing that a new day will come, but realizing that this one is just fine with me.

What a good time it is to be in this constellation: winter and snowing, middle Midwest, late in life, and home. I am blessed.

On *Writer's Almanac* this morning Garrison Keillor read a poem by Henry Wadsworth Longfellow, "Snow-Flakes." The poem begins:

Out of the bosom of the Air,
    Out of the cloud-folds of her garments shaken,
Over the woodlands brown and bare,
    Over the harvest-fields forsaken,
        Silent, and soft, and slow
        Descends the snow.

The skies continue to shake snow and cover the fields below.

A natural melancholy, a creative force, then as now, known by Longfellow and known by us.

This is the poem of the air,
    Slowly in silent syllables recorded;
This is the secret of despair,
    Long in its cloudy bosom hoarded,
        Now whispered and revealed
        To wood and field.

Accompanied by the photographs of the day's landscape. Of the horizon where sky meets wood and field. Terrestrial sightings made precious by the singularity of happening, only seen once like this.

Emerson's words from his essay "Nature" are my guide and inspiration this day. "To the attentive eye, each moment of the year has its own beauty, and in the same field, it beholds, every hour, a picture which was never seen before, and which shall never be seen again." The passage ends in a sentiment shared with Longfellow: "The heavens change every moment, and reflect their glory or gloom on the plains beneath." Each photograph is an instant of something once seen by an eye of certain mind and spirit.

What are we to make of the report that the remains of Jesus and his family, including those of Mary Magdalene, have been discovered in a crypt in Jerusalem? We wait with great interest for further information. A host on local radio this morning, in jest, said that a box containing the remains of Jesus' Cousin Raul was also found in the crypt. The bodily resurrection of the historical Jesus, the one known as Christ, is now surely in doubt. Even if the discovery is false, a core religious doctrine has been questioned. With or without resurrection, we all face the inevitable ending of natural life on earth. A Christian belief in resurrection might calm the fear of death, but is offset by the fear of judgment.

Still reading the poems of Longfellow, prompted by the discovery of my grandfather's grade school *Fifth Reader,* I have found Longfellow's poem titled "Nature." The first line has the mother leading her child by the hand to bed. The last lines of the poem turn to last things, to the diminishment of life and the ending of life. Asking for a gentle ending to life:

> So Nature deals with us, and takes away
> Our playthings one by one, and by the hand
> Leads us to rest so gently, that we go
> Scarce knowing if we wished to go or stay,
> Being too full of sleep to understand
> How far the unknown transcends the what we know.

Too tired, and ill, to understand even how far we are from knowing the unknown. An unknown that might be little more than oblivion.

I'll listen once more to Billy Joe Shaver's song "Ride Me Down Easy," recorded in 1973 by Bobby Bare. "So won't you ride me down easy Lord, ride me on down. Leave word in the dust where I lay. Say, 'I'm easy come, easy go and easy to love when I stay.'" That would be grace.

One would hesitate not to include Shakespeare's Sonnet 73 at the end of a book such as this one. A rounding, from beginning to end, through the seasons of a life in this one world. Leaves have weathered and boughs shake against the cold.

Against the bare ruined choir of the abandoned church "where late the sweet birds sang."

> That time of year thou mayst in me behold
> When yellow leaves, or none, or few, do hang
> Upon those boughs which shake against the cold,
> Bare ruined choirs, where late the sweet birds sang.
> In me thou seest the twilight of such day
> As after sunset fadeth in the west,
> Which by and by black night doth take away,
> Death's second self that seals up all in rest.
> In me thou seest the glowing of such fire
> That on the ashes of his youth doth lie,
> As the death-bed whereon it must expire,
> Consumed with that which it was nourished by.
> This thou perceiv'st, which makes thy love more strong,
> To love that well which thou must leave ere long.

Death comes from the ashes that once nourished this life. But still, there is a glowing of the flame. Shakespeare's aging man is more than the ashes. An ending, to be certain, but in the glowing rather than the fading. A love made even stronger by the leaving of this world. And in the ending of the winter, as the signs of spring appear, you long and love even more.

Harry Callahan, the photographer who continues to provide me with inspiration and ideas for photographing, was interviewed late in his life. In one interview he reflected on his art and spiritual life. "I felt that I wanted to do something that would benefit humanity. I wanted something important, something spiritual in my life. And as these feelings developed,

I found that my spiritual enrichment came through art. It was then that I realized I wanted to be an artist. That's it. You see, I learn slowly, but every day I seem to learn more."

A couple of years later, he told an interviewer that even in old age he wanted to express himself in his photographs, in his photographing on the streets and in the parks of Atlanta where he was now living. He knew that he could not go jogging, but he hoped that his life at this age could be expressed in his art. He said, "So far I still look forward to going out and photographing." The flame, the fancy, the knowing of the preciousness of this life on earth.

Looking now and then above the horizon, beyond the grounding of this earthbound life. You look through the treetops into the sky above.

You can love nature fully only in the unconditional acceptance of nature's ways: birth, decay, death, and transformation. Recognition of the mortality of everything that lives, animal and vegetable. Knowing the reality of impermanence, in spite of a lifetime of human conditioning to the contrary. Such understanding is necessary for the reverence of nature. I look into the sky, through the branches of the trees, in training for reverence.

Thoreau, in living one world at a time, knew that in this world nature is not necessarily a benevolent force. This became evident to him as he walked the shores of Cape Cod and looked out to the vast and tumultuous ocean. Venturing from his home in Concord in the fall of 1849, he arrived just as Irish immigrants were washing up on the shore at Cohasset from the shipwreck in the stormy sea. Paul Theroux reminds us in his introduction to a recent edition of Thoreau's *Cape Cod* that Thoreau's wanderings on Cape Cod led to a personal discovery of the ocean beyond land and shore. After his weeks of being there, Thoreau would write these poetic lines about his life: "My life is like a stroll upon the beach / As near the ocean's edge as I can go." The poem ends: "Along the shore my hand is on the pulse, / And I converse with many a shipwrecked crew."

Yet there was a beauty to be seen on these shores. A beauty, he wrote, "enhanced by wrecks like this, and it acquired thus a rarer and sublime beauty still." Thoreau, a landlubber, had become a beachcomber and gazer at the sea. With back to the land, standing on the shore, he watches the ocean, strange and dark. Beyond the wreckage true nature is revealed: "The sea, vast and wild as it is, bears thus the waste and wrecks of human art to its remotest shore. There is no telling what it may not vomit up. It lets nothing lie."

This earth on which we live is but a shore. A shore that looks out to the open sea. The sea that Paul Theroux reminds us is a wilderness. No amount of travel inland by Thoreau could compare to the wilderness of the ocean beyond the shore. Thoreau would write: "The ocean is a wilderness reaching round the globe, wilder than a Bengal jungle, and fuller of monsters." There is a sense of release, of freedom, in this vision.

I will go now out to the oak savanna beside the lake only a few blocks from my house. Standing below the tall oaks, camera in hand, I will gaze through the branches into the vast sky above. The sky, for this inlander, is twin to the sea. Behold, the wildness of the infinite. Standing on this earth, in the one world in which I have lived, I am in awe of nature and its unfathomable meaning. A wilderness entertained sporadically and carefully with reverence.

A beautifully published book of a selection of poems by Robert Frost, containing black and white photographs, arrived in the morning mail, bearing the title *Versed in Country Things*. The photographer, B. A. King, notes that his photographs cannot appropriately illustrate the poems, but they can serve as acceptable neighbors to the poems.

Preceding Frost's poem "Birches" are four haunting photographs of farmers, families, and their horses working in the fields. In the poem, Frost talks of the farm boy who, between times of work in the fields, plays alone, swinging on the branches of the birch trees, bending branches to the ground. Bending branches as ice storms do, but with a difference not found in the nature of the storm.

> When I see birches bend to left and right
> Across the lines of straighter darker trees
> I like to think some boy's been swinging them.
> But swinging doesn't bend them down to stay
> As ice storms do.

Frost as narrator of the poem would prefer the farm boy's play over the other forces of nature.

> But I was going to say when Truth broke in
> With all her matter of fact about the ice storm,
> I should prefer to have some boy bend them
> As he went out and in to fetch the cows—
> Some boy too far from town to learn baseball,
> Whose only play was what he found himself,
> Summer or winter, and could play alone.

The boy swinging, "with a swish / Kicking his way down through the air to the ground." And Frost writing, "So was I once myself a swinger of birches."

Later in life, the narrator of the poem dreams of going back to the birches, "when I'm wary of considerations." Swinging on the branches, and coming back to earth again. "I'd like to get away from earth awhile / And then come back to it and begin over." With the wonderful lines that end the poem:

> Earth's the right place for love:
> I don't know where it's likely to go better.
> I'd like to go by climbing a birch tree,
> And climb black branches up a snow-white trunk
> *Toward* heaven, till the tree could bear no more,
> But dipped its top and set me down again.
> That would be good both going and coming back.
> One could do worse than be a swinger of birches.

You wish to be forever a swinger of birches. This is the nature for all of us versed in human things. Grounded as we are, here on earth, under the vast and infinite sky above.

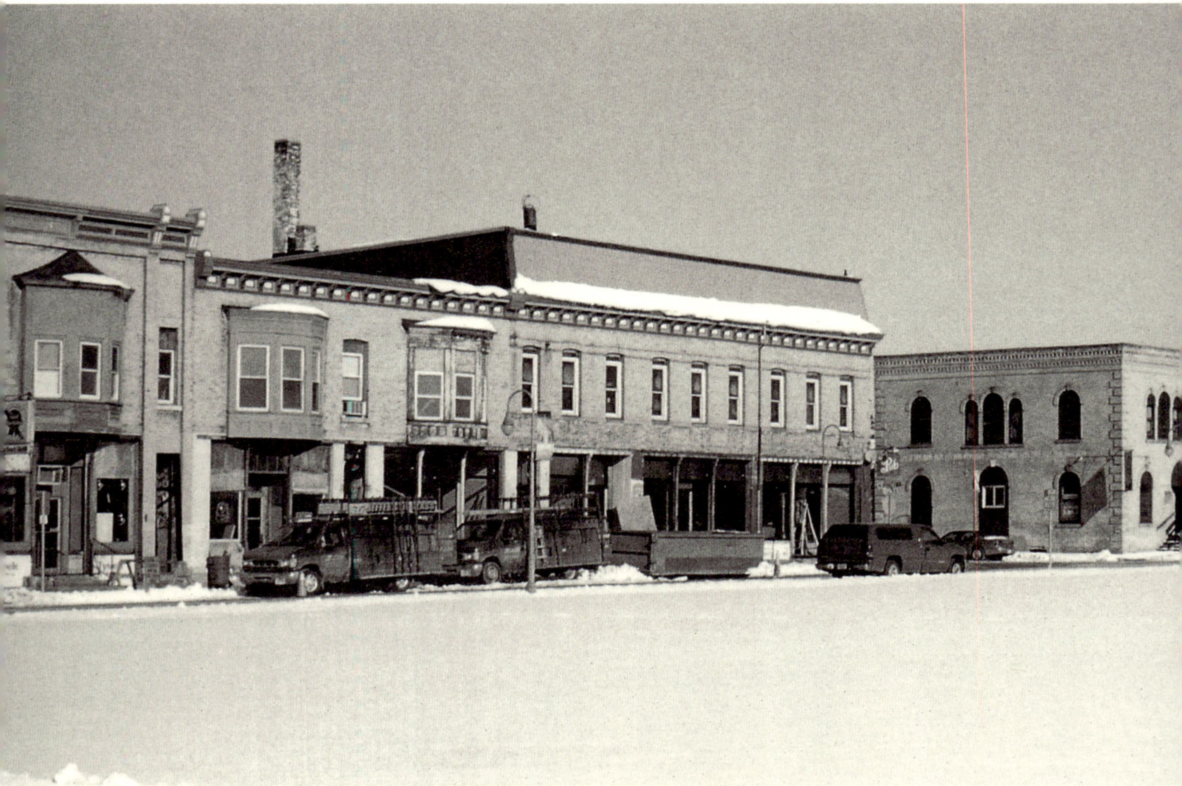

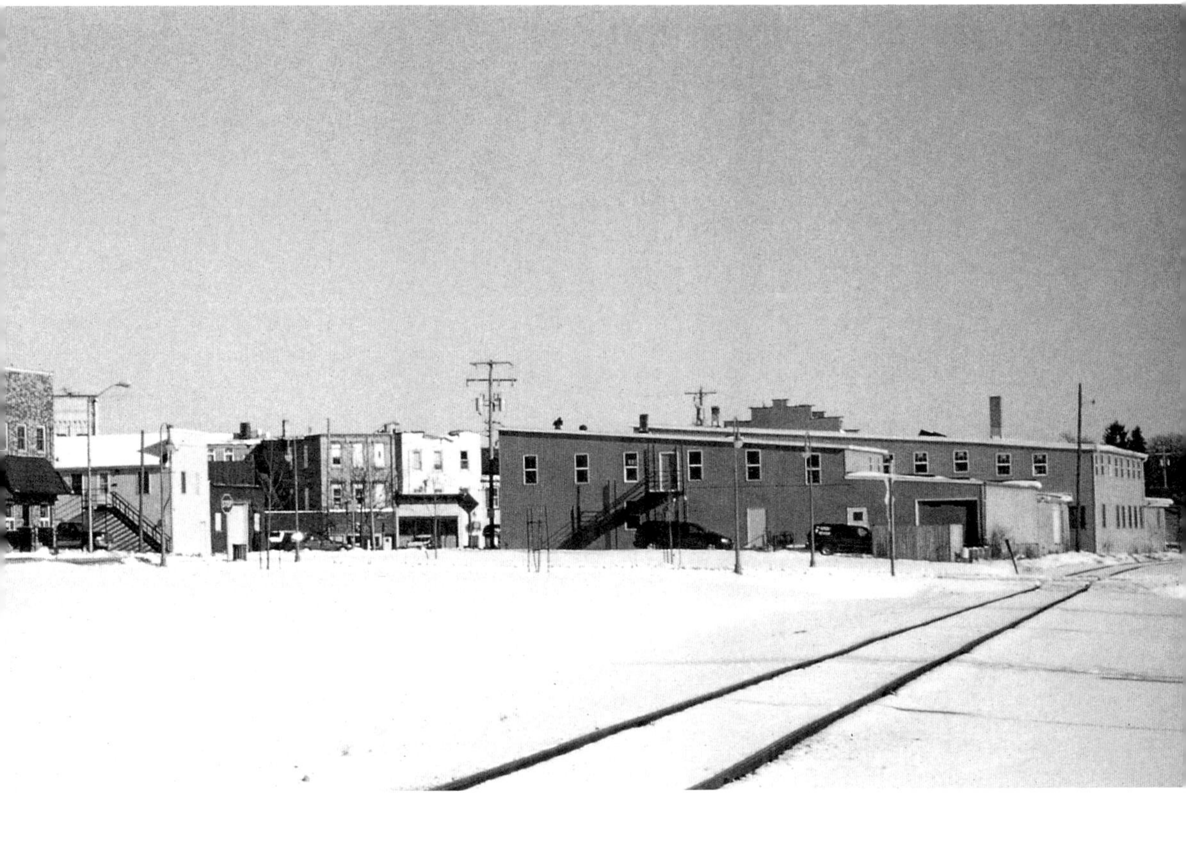

Bibliography

Ackroyd, Peter. *J. M. W. Turner.* New York: Doubleday, 2005.

Aiken, Robert. *The Mind of Clover: Essays in Zen Buddhist Ethics.* San Francisco: North Point Press, 1984.

Aurelius, Marcus. *Meditations.* New York: Alfred A. Knopf, 1992.

Babcock, Harold. "The Sacred in Images." *UU World,* Winter 2005, 30–35.

Berry, Wendell. *The Selected Poems of Wendell Berry.* Washington, DC: Counterpoint, 1999.

The Bhagavad Gita. Trans. Eknath Easwaran. Petaluma, CA: Nilgiri Press, 1985.

Bloom, Harold. *American Religious Poems: An Anthology.* Ed. Harold Bloom and Jesse Zuba. New York: Library of America, 2006.

———. *Jesus and Yahweh: The Names Divine.* New York: Riverhead Books, 2005.

Budbill, David. *While We've Still Got Feet.* Port Townsend, WA: Copper Canyon Press, 2005.

Bukkyo Dendo Kyokai. *The Teaching of Buddha.* Tokyo: Buddhist Promoting Foundation, 1966.

Bussan, Paul. *A Rage of Intelligence.* Hamden, CT: PSB Publishing, 2003.

Camus, Albert. *The Myth of Sisyphus.* Trans. Justin O'Brien. New York: Alfred A. Knopf, 1955.

———. *The Stranger.* 1946. Trans. Matthew Ward. New York: Alfred A. Knopf, 1993.

Cassill, R. V. *Writing Fiction.* Saddle River, NJ: Prentice Hall, 1975.

Chah, Achaan. *A Still Forest Pool.* Ed. Jack Kornfield and Paul Breiter. Wheaton, IL: Theosophical Publishing House, 1985.

Chekhov, Anton. *The Essential Plays.* Trans. Michael Henry Heim. New York: Modern Library, 2003.

Creeley, Robert. *On Earth.* Berkeley: University of California Press, 2006.

Crosby, Donald A. *A Religion of Nature*. Albany: State University of New York Press, 2002.

Dalai Lama. *The Universe in a Single Atom: The Convergence of Science and Spirituality*. New York: Morgan Road Books, 2005.

Dalai Lama and Howard C. Cutler. *The Art of Happiness: A Handbook for Living*. New York: Riverhead Books, 1998.

Davis, Charles F. *Harvest of a Quiet Eye: The Natural World of John Burroughs*. Madison, WI: Tamarack Press, 1976.

The Dhammapada. Trans. Irving Babbitt. New York: New Directions, 1936.

Dott, Robert H. Jr., and John W. Attig. *Roadside Geology of Wisconsin*. Missoula: Mountain Press Publishing, 2004.

Emerson, Ralph Waldo. *Nature and Other Writings*. Boston: Shambhala, 1994.

Fowles, John, and Frank Horvat. *The Tree*. Boston: Little, Brown, 1979.

Franck, Frederick. *The Zen of Seeing*. New York: Vintage Books, 1973.

Frost, Robert. *Versed in Country Things*. Photographs by B. A. King. Poems selected by Edward Connery Lathem. Boston: Little, Brown, 1996.

Gopnik, Adam. "Rewriting Nature." *New Yorker*, October 23, 2006, 52–59.

Greenough, Sarah. *Harry Callahan*. Washington, DC: National Gallery of Art, 1996.

Hagen, Steve. *Buddhism Is Not What You Think*. New York: HarperSanFrancisco, 2003.

Hamill, Sam, and J. P. Seaton, trans. and ed. *The Poetry of Zen*. Boston: Shambhala, 2004.

Heaney, Seamus. *Finders Keepers: Selected Prose, 1971–2001*. New York: Farrar, Straus and Giroux, 2002.

Hoagland, Edward. *Compass Points: How I Lived*. New York: Pantheon Books, 2001.

———. *Hoagland on Nature: Essays*. Guilford, CT: Lyons Press, 2003.

Jefferies, Richard. *The Story of My Heart: My Autobiography*. 1883. Totnes, UK: Green Books, 2002.

Jeffers, Robinson. *Rock and Hawk*. Ed. Robert Hass. New York: Random House, 1987.

Kafka, Franz. *The Castle*. New York: Alfred A. Knopf, 1992.

Kasser, Rodolphe, Marvin Meyer, and Gregor Wurst, eds. *The Lost Gospel of Judas*. Washington, DC: National Geographic Society, 2006.

Kimmelman, Michael. *The Accidental Masterpiece: On the Art of Life and Vice Versa*. New York: Penguin Press, 2005.

King, Martin Luther Jr. *I Have a Dream*. New York: HarperSanFrancisco, 1992

Klinkenborg, Verlyn. *Timothy; or, Notes of an Abject Reptile*. New York: Alfred A. Knopf, 2006.

Lozoff, Bo. *We're All Doing Time.* Durham, NC: Hanuman Foundation, 1985.
Mabey, Richard. *Gilbert White: A Biography of the Author of* The Natural History of Selborne. London: Century Hutchinson, 1986.
Merton, Robert K. *Sociological Ambivalence.* New York: Free Press, 1976.
Merton, Thomas. *The Way of Chuang Tzu.* New York: New Directions, 1960.
Murphy, Susan. "A Promise to the Earth: The Practice of Place." *Inquiring Mind,* Spring 2006, 18–19.
Naipaul, V. S. *Half a Life.* New York: Alfred A. Knopf, 2001.
———. *Literary Occasions.* New York: Alfred A. Knopf, 2003.
The New Oxford Annotated Bible. New Revised Standard Version. Ed. B. M. Metzger and R. E. Murphy. New York: Oxford University Press, 1991.
Nhat Hanh, Thich. *Be Free Where You Are.* Berkeley: Parallax Press, 2002.
O'Faolain, Sean. *The Irish.* Hamondsworth, UK: Penguin, 1947.
Peattie, Donald Culross. *A Natural History of Trees of Eastern and Central North America.* 1964. Boston: Houghton Mifflin, 1991.
Richardson, Robert D. Jr. *Emerson: The Mind on Fire.* Berkeley: University of California Press, 1995.
———. *Henry Thoreau: A Life of the Mind.* Berkeley: University of California Press, 1986.
———. *William James: In the Maelstrom of American Modernism.* Boston: Houghton Mifflin, 2006.
Schmidt, Leigh Eric. *Restless Souls: The Making of American Spirituality.* New York: HarperSanFrancisco, 2005.
Sennett, Richard, and Jonathan Cobb. *The Hidden Injuries of Class.* New York: Alfred A. Knopf, 1972.
Sheldons' Modern School Fifth Reader. New York: Sheldon & Company, 1885.
Sontag, Susan. *Regarding the Pain of Others.* New York: Farrar, Straus and Giroux, 2003.
Stafford, William. *The Way It Is: New & Selected Poems.* St. Paul: Graywolf Press, 1998.
Stegner, Wallace. *Where the Bluebird Sings to the Lemonade Springs.* New York: Random House, 1992.
Sullivan, Dennis. "The Political Economy of Eternal Life." *Justitia,* Winter 2007, 1–5.
———. "Richard Quinney on the Transformation of Self and Others: An Interview." *Contemporary Justice Review* 9 (September 2006): 277–82.
Tao Te Ching. Trans. Stephen Mitchell. New York: Harper & Row, 1988.
Taylor, Ian, Paul Walton, and Jock Young. *The New Criminology.* London: Routledge & Kegan Paul, 1973.
Thoreau, Henry David. *Cape Cod.* Introd. Paul Theroux. New York: Viking Penguin, 1987.

———. *Walden*. Ed. J. Lyndon Shanley. Princeton: Princeton University Press, 1973.

———. *Walden and Resistance to Civil Government*. New York: W. W. Norton, 1992.

The Upanishads. Trans. Eknath Easwaran. Petaluma, CA: Nilgiri Press, 1987.

Vaughan, William, Elizabeth E. Barker, and Colin Harrison. *Samuel Palmer, 1805–1881: Vision and Landscape*. Burlington, VT: Lund Humphries, 2005.

Vender, Helen. *The Art of Shakespeare's Sonnets*. Cambridge, MA: Harvard University Press, 1997.

Watts. Alan W. *The Way of Zen*. New York: Random House, 1957.

Weschler, Lawrence. *Vermeer in Bosnia*. New York: Pantheon Books, 2004.

White, Gilbert. *The Natural History of Selborne*. 1789. Ed. Richard Mabey. London: Century Hutchinson, 1988.

Whitehead, Alfred North. *Modes of Thought*. New York: Capricorn Books, 1958.

Whitman, Walt. *Complete Poetry and Collected Prose*. New York: Library of America, 1982.

Wordsworth, William. *The Poems of William Wordsworth*. Ed. Jonathan Wordsworth. Cambridge, UK: University Printing House, 1973.

Yeats, W. B. *Mythologies*. New York: Collier, 1959.

Young, Neil. "Prairie Wind." CD. Burbank, CA: Reprise Records, 2005.

Photographs by Richard Quinney: p. vi, *Madison, Wisconsin;* p. 10, *East of the Barn, Quinney Farm;* p. 36, *Quinney Farm from the North;* p. 70, *Frank Lloyd Wright's Taliesin, Spring Green, Wisconsin;* p. 92, *Ice Fishing on Monona Bay, Madison, Wisconsin;* p. 124, *Whitewater, Wisconsin*

Colophon
This book is set in Sabon, an old-style serif typeface designed by the German-born typographer and designer, Jan Tschichold (1902–1974). It was designed by Ken Crocker. Capital Offset Company Inc. printed the book, and it was bound by Acme Bookbinding.